Lecture Notes in
Computer Science

T0219993

Lecture Notes in Computer Science

Lecture Notes in Computer Science

Edited by G. Goos and J. Hartmanis

102

James Harold Davenport

On the Integration of Algebraic Functions

Springer-Verlag
Berlin Heidelberg New York 1981

Author

James Harold Davenport
Emmanuel College
Cambridge
England

AMS Subject Classifications (1979):
CR Subject Classifications (1979): 5.25, 5.7

ISBN 3-540-10290-6 Springer-Verlag Berlin Heidelberg New York
ISBN 0-387-10290-6 Springer-Verlag New York Heidelberg Berlin

Acknowledgements

I wish to express my gratitude to all the people who have made this work possible. The programming could not have been done without the expert advice of Drs. J.P. ffitch and A.C. Norman on their LISP system, and the advice of Prof. A.C. Hearn, Dr. A.C. Norman and Mr. R.G. Hall on the REDUCE-2 system. I am also grateful to Mrs. P.M.A. Moore for her permission to incorporate several of her changes to REDUCE-2.

I am grateful to the Director and staff of the Cambridge University Computing Service, where nearly all the work described here was performed, for their support of my apparently endless computations and for their advice, especially that of Mr M.J.T. Guy and Mr. C.E. Thompson. The implementation of the integration system at the IBM Thomas J. Watson Research Centre relied on the advice of Mr. J.E. Harry, and the MTS implementation was performed with the assistance of Mr. W.A. Dodge and Mrs. J. Caviness of Rensselaer Polytechnic Institute and Mr. M. Alexander of the University of Michigan. Many improvements to the factoriser were made by Dr. D. Dahm (Burroughs Corp.), who also pointed out several other bugs. Many problems with the LISP interface were discovered and corrected by Mr. D. Morrison (University of Utah).

I am very grateful to the many mathematicians whose advice I have sought, especially Prof. Sir Peter Swinnerton-Dyer, Dr. B.J. Birch, Prof. D.J. Lewis, Prof. S. Maclane, Dr. R.H. Risch, Prof. A. Schinzel and Prof. M.F. Singer. My original introduction to computers and algebraic geometry was performed by Dr. N.M. Stephens. The discussion in chapter 6 on special values of parameters was inspired by discussion with Dr. P.M. Neumann and the Oxford University Number Theory Seminar.

I would also like to acknowledge the help I have received from many discussions with Dr. J.P. ffitch, Prof. B.F. Caviness and the Rensselaer Polytechnic Institute seminar, Dr. R.D. Jenks and Dr. D.Y.Y. Yun of the IBM Thomas J. Watson Research Centre, and the MACSYMA group at M.I.T. (especially Mr. B.M. Trager).

I am very grateful to Professor Sir Peter Swinnerton-Dyer and Professor A. C. Hearn, who read an earlier version of this work, for their many helpful comments and suggestions.

Professor M.F. Singer also read a version and made many valuable remarks. Earlier versions of this work were prepared at the Cambridge University Computing Service, and I am grateful to Dr. A.J. Herbert and Mr. M.A. Johnson for their advice on text processing. This version was prepared at the IBM Thomas J. Watson Research Centre using the Yorktown Formatting Language, and I am grateful to the text processing consultants, Miss A.M. Gruhn, Miss K.C. Keene and Mrs. C.H. Thompson, for their advice.

Finally I would like to thank my mother for her assistance, especially with the references, and Dr. A.C. Norman, my supervisor, for his continued advice and encouragement.

Contents

Contents

Introduction

This work is concerned with the following question: *"When is an algebraic function integrable?"*. We can state this question in another form which makes clearer our interpretation of integration: "If we are given an algebraic function, when can we find an expression in terms of algebraics, logarithms and exponentials whose derivative is the given function, and what is that expression?".

This question can be looked at purely mathematically, as a question in decidability theory, but our interest in this question is more practical and springs from the requirements of computer algebra. Thus our goal is *"Write a program which, when given an algebraic function,* will produce an expression for its integral in terms of algebraics, exponentials and logarithms, or will prove that there is no such expression".

Computer Integration in General

In this section[*] I discuss briefly the whole area of computer indefinite integration, and explain the relationship of the problem posed above to computer algebra in general.

The simplest class of functions which we may wish to integrate are the polynomials. Here the solution is extremely simple: all polynomials are integrable, and the algorithm is merely to replace aX^n by $aX^{n+1}/(n + 1)$.

The next case comprises the rational functions, and here again every rational function is integrable. However, the solution is not as straightforward as in the previous paragraph.

---

[*] This section is largely based on a recent survey of the computer integration field (Norman & Davenport, 1979). I am grateful to Dr. Norman for many helpful discussions.

In the first place the integral of a rational function need not be a rational function, but can include certain logarithms. In the second place, we may need to extend the constant field in order to integrate rational functions (see Risch, 1969, Proposition 1.1 for a simple example of this). As a consequence of this, while there have been algorithms for integrating rational functions since Liouville and Hermite, there is still scope for improvement, and the latest progress is contained in work by Trager (1976) and Yun (1977b). This area has also stimulated much research into the handling of algebraic numbers (Trager, 1976 and Zippel, 1977).

The obvious extension of rational functions is to algebraic functions: see the section on "Previous Work" below. This is the first case where there are non-integrable functions, and this is a major reason why the algebraic case is so much more difficult than the rational case.

The other extension of rational functions is to pure transcendental functions, i.e. functions lying in a purely transcendental tower of fields based on K(x) (assuming K to be a field of constants, and x the variable of integration) where each extension is generated purely transcendentally[#] over the previous field by either the logarithm or the exponential of a member of that field. This case of the integration problem was settled completely by Risch (1969), whose algorithms are largely implemented in SIN (Moses, 1971). A different approach has been taken by Risch & Norman (Norman & Moore, 1977) and by Rothstein (1977).

The work described in this monograph, which produces a complete algorithm for the algebraic case, means that there are now algorithms for integrating both purely algebraic and purely transcendental functions. We will discuss in chapter 9 the possibility of integrating functions which are neither (i.e. transcendental but not purely transcendental).

E.g. we may not construct $\exp(\log(1+x)/2)$ since this is $(1 + x)^{1/2}$ and is therefore algebraic rather than transcendental.

Outline of the Book (1)

The production of a computer program to find the integrals of algebraic functions or prove them unintegrable involves both pure mathematics and computer science. We can produce a partial algorithm* for the problem of integrating algebraic functions without the use of too many advanced concepts (Chapters 2-4), but converting this into a complete algorithm requires substantially more mathematics (chapters 5-8).

Before we can describe the material covered in the monograph in more specific terms, we must understand more of the overall structure. This is essentially a vicious circle, but one that can be broken by treating a rational function as though it were an algebraic function, as is done in the next section.

Illustration

The methods we will use for integrating algebraic functions are based on algebraic geometry and on number theory, so it is difficult to describe an example before the underlying theory has been explained. Therefore we will consider a rational function and integrate it by the methods that would be used for algebraic functions. This has the advantage that most of the sub-algorithms reduce to triviality.

Consider the problem of integrating $dx/(x-1)(x-2)^2$. If $y = (x-1)$, then we can write the integrand as $y^{-1}(1 + 2y + 3y^2 + ...)dy$, and integrating this term-by-term implies that there is a $\log y$ term in the answer. Writing $z = (x-2)$ we express the integral as $z^{-2}(1-z + z^2 + ...)dz$, and integrating this term-by-term we deduce that there is a $\log z$ term and a $-1/z$ term.

Hence the integral contains $\log (x-1) - \log (x-2) - 1/(x-2)$, and is in fact precisely

---

* I.e. an algorithm which will terminate with the correct integral if the integrand is elementarily integrable, but which may run for ever if it is not (rather than replying that it is not).

this.

We have glossed over one very important point: what about the value $x = $ infinity. Might we not need a term from the power series expansion about there? Writing $z = 1/x$, we wish to express the integrand as a power series in z:

$$\frac{dx}{(x-1)(x-2)^2} = \frac{z^3 dx}{(1-z)(1-2z)^2} = \frac{z^3(-dz/z^2)}{(1-z)(1-2z)^2} = \frac{-z\,dz}{(1-z)(1-2z)^2},$$

so there is no logarithmic term coming from $x = $ infinity.

Outline of the Book (2)

In the previous example, we expanded the integrand in power series, deduced logarithmic terms from residues (i.e. the coefficient of z^{-1} in a power series expansion in terms of z), and deduced the integral from the logarithmic part and the rest of the power series expansions. These 3 steps correspond to the three chapters (2,3,4) providing a partial algorithm for the integration of algebraic functions.

Chapter 2 discusses the representation of algebraic functions in computer algebra, the question of *places* in algebraic geometry[#] and the question of Puiseux expansions (which are the correct equivalent of the power series expansions which we were using). Chapter 3 discusses Coates' Algorithm, which finds a function with poles and zeros as specified. This turns out to be needed not only for finding the arguments of logarithms, but also for evaluating the algebraic part and for many other tasks in the integration algorithm. Chapter 4 states and proves Risch's Theorem, which (subject to matters already outlined) reduces the problem of integrating algebraic functions to the *torsion divisor* problem. In particular, if we assume that all divisors could be torsion divisors, we have the partial algorithm mentioned above.

[#] "Place" is the appropriate generalisation of the concept of "value of x" which we used when we expanded about $x = 1$ or $x = 2$ (or even infinity).

Chapters 5-8 of the book discuss the torsion divisor problem, which can be (approximately) rephrased as: *"given a list of places at which we require the argument of a logarithm to have certain poles or zeros, is there a function which has these poles and zeros, but to a constant multiple of the required degree?"* (e.g. we may not be able to find a function with poles and zeros of order 1, but we may be able to find one with poles and zeros of order 3). In this case the cube root of that function would in fact have poles and zeros of order 1. Since $\log (f(x)^{1/3}) = \log (f(x))/3$ all we have to do is adjust the coefficient of the logarithm.

Chapter 5 discusses the general theory of this problem, and provides answers in certain special cases. In chapter 6 we consider the problem when there is an independent transcendental parameter. In this case there is a complete solution which allows us either to state whether or not the divisor is a torsion divisor without our needing to find the degree of torsion, or to reduce the torsion problem to one not involving the parameter (which problems are considered in the later chapters). If the integral is not elementary, we can then ask the question: *"For what values of the parameter is the integral elementary"* and this question is also discussed in chapter 6.

Chapter 7 discusses the other possibility (i.e. there are no transcendental parameters, so that the problem is defined over an algebraic number field) in the case of elliptic curves. In this case we can apply the highly developed mathematical theory of elliptic curves to determine whether the divisor is a torsion divisor, and if so what the degree of torsion is. Chapter 8 discusses the remaining case (i.e. algebraic number fields when the curves are not elliptic) and produces an algorithm for finding an bound for the torsion of the divisor, so that we can apply Coates' Algorithm to each multiple of the divisor up to this bound, and then, if we had not found a multiple for which Coates' Algorithm yields a corresponding function, we can declare that the original function does not have an elementary integral.

Theoretical Limitations

One serious difficulty with the problem as formulated is that "logarithm" is a many-valued function. Following Risch (1969) we can eliminate this difficulty by taking a purely algebraic approach to the problem, i.e. by defining log u to be that function f such that f' $= u'/u$: interpreting this as a definition in the theory of differential algebra (Kolchin, 1973, see also Ritt, 1948).

There is a more fundamental difficulty in the posing of the problem. There exist unintegrable algebraic functions $f(x)$ (see Appendix 2 Example 2), and so, if c is any constant, we know that $cf(x)$ is integrable if and only if $c = 0$. But it is known (Richardson, 1968), that there can be no algorithm for deciding if an expression generated from the integers, log 2 and π by the operations of addition, subtraction, multiplication, division and the taking of exponentials, logarithms and absolute values is zero or not. Putting these results together shows that there is a sense in which the integration problem[*] is formally undecidable. It is nevertheless clear that the undecidability which we have demonstrated above is somewhat spurious, since it depends on the properties of numbers and not on those of integrals.

We will circumvent this restriction by requiring all algebraic dependencies between constants in the integrand to be explicitly stated. The constants of the integrand have then to be rational numbers, algebraically independent transcendentals or expressions which are explicitly algebraically dependent on such expressions.

One practical illustration of the necessity for making all algebraic relationships explicit can be seen in our implementation of Manin's work (Chapter 6), where we require to differentiate with respect to a transcendental constant. This will only work if we know precisely which other constants are algebraically dependent on the one in question.

---

[*] Note that we cannot use the theorem (Richardson, 1968) about the undecidability of the integration problem directly, because that theorem relies on the ability to take logarithms and absolute values in the function field, and we only allow these operations in the constant domain.

As pointed out in Chapter 6, our requirement for explicit relationships means that we cannot guarantee[#] to produce correct results when working with integrals involving both e and π explicitly, since we do not know if the two constants are algebraically independent or not.

Previous Work

The first significant attempt to integrate algebraic functions by computer program was the SAINT system (Slagle, 1961). This was a heuristic system for integrating functions based on a variety of transformations and simplifications. It was extremely successful in solving simple problems, but its heuristics were all aimed at finding integrals: it did not address the problem of proving expressions to be unintegrable.

The next major development in the integration of algebraic functions (and indeed in computerised integration generally) was the SIN system (Moses, 1967 and 1971). This incorporated the Risch (1969) algorithm for purely transcendental functions, as well as a variety of special transformation and simplification rules. This gave a program that is very useful in practice and which can integrate most "ordinary" algebraic functions, although it does not contain any deep theory and cannot be generalised to cover harder integrals (e.g. Appendix 2, Examples 3-5).

While its treatment of algebraic functions was not algorithmic, it was in certain cases capable of reducing an algebraic function to a purely transcendental function, where it could apply an algorithm (for example it would substitute $y = \cosh(x)$ in $\sqrt{x^2-1}$ to make the integrand transcendental). However, this form of rationalising is normally only possible for integrands over curves of genus 0 (see Chapter 2) and all these functions are integrable, so the partially algorithmic nature of this program is not as useful as might

---

[#] But note that if we assume that the two constants are algebraically independent, and produce an integral, then this is still valid if the constants are in fact algebraically dependent. It is only demonstrations of unintegrability that break down.

appear, since the main practical advantage of algorithmic methods is that they can prove unintegrability.

Another significant advance in the field was made by Ng (1974), building on the mathematical advances of Carlson. By using the theory of R-functions, Ng was able to solve completely the question of elliptic and hyper-elliptic integrals (i.e. those containing just one square root).

More recently Trager has been working in this area (Trager, 1978). His (unpublished) results generalise earlier work (Trager, 1976) on the constant field of definition for rational integrals. However there is currently no implementation of these significant results. More recently (Trager, 1979) he has been working on the special case of *simple radical extensions* (by this we mean that the only algebraic quantity present depending on the variable of integration is the n-th root (for some n) of a rational quantity) and he has produced some very effective algorithms for finding the algebraic part of integrals in such extensions, which appear to implement some of the ideas of Chebyshev (1853).

Timings

We will frequently quote specific examples of integrals and state how long the implementation of the algorithm takes to find the integral or discover that they are unintegrable. Unless otherwise stated, all such times will refer to the implementation on the IBM 370/165[$] at Cambridge University running a modified version of OS/MVT release 21.6. The integration algorithm is implemented as a part of REDUCE-2 (Hearn, 1973) version of March 6th., 1978 (with several modifications: see Appendix 1 for a description of the more significant ones), and runs under the Cambridge version of LISP

[$] This computer has been fitted with faster main memory than is usual. The net effect of this change is hard to quantify, but on the sort of programs we are running the CPU time is about 10% less than on an unmodified 370/165. The DAT (Dynamic Address Translation, i.e. virtual memory) feature is not installed.

ffitch & Norman, 1977). The times quoted are CPU times, and exclude garbage collection* time. Too much should not be read into details of the timings since they are inherently inaccurate to the extent of about 15% because of inter-run variations.

These times are of little value except for comparison with similar integrals run on the same version of the program because of the extent to which they depend on the implementation of sub-algorithms, not only those described in this monograph, but also the fundamental algorithms of computer algebra as supplied in the REDUCE-2 system. For instance the modification to the greatest common divisor routine described in Appendix 1 item 3 cut the times for some integrals by a factor of at least 10. Also the time for an integral can vary greatly for reasons not connected with its intrinsic difficulty: there is a set of illustrations of this in Appendix 2 example 1. The times quoted do not give a completely fair impression of the speed of the program on comparatively simple cases since they were run on the development version of the program, which contains extra checks and printing and also does not contain many "short-cuts" of great practical importance, but no theoretical significance. There is an example of the difference that can be made by the use of a "production" version in Appendix 2 Example 6.

---

* This is usually about 30% of the non-garbage collection time if the program is running in sufficient store (say 700Kbytes for most examples). This time also includes the time required to load compiled LISP functions from disc, and this can represent as much as 2 seconds for complicated cases.

Algebraic Computations

Algebraic Relationships

Algebraic relationships between variables or expressions are very common in computer algebra. Not only do they often occur explicitly, in forms like SQRT(X^2+1), but well known difficulties such as sin(x)**2 + cos(x)**2 = 1 (Stoutemyer, 1977) can be expressed in this form. Nevertheless it is difficult to compute with regard to these algebraic relationships. This chapter discusses the problem of such computing, and then enters the area of algebraic geometry, which is a natural outgrowth of attempts to perform such computations as readily as one computes without them.

The simplest case of computing subject to algebraic relationships consists of computations in the field $\{K(X,Y) \mid F(X,Y) = 0\}$ where F is a rational function (which can be reduced to a polynomial by multiplying out any denominator) in which X and Y appear effectively, and K is a field of characteristic* 0. As a simple example, I shall consider $F(X,Y) = Y^2-X^2 + 1$ as the definition of the algebraic relationship.

One of the key ideas in efficient computer algebra is that of *canonical forms,* by which is meant ensuring that equivalent expressions, however computed, are represented by equivalent data structures. This has the great merit that the case of unequal expressions is normally detected very early in any comparison operation. Since our computations are likely to tax the power of even the largest computers, this point is one we cannot afford to neglect.

The first problem that arises in the example is that Y^2 and X^2-1 are the same

---

* We will not normally (except in Chapter 8) be concerned with fields of finite characteristic, and any reference to a field should be construed as a reference to a field of characteristic 0 unless the converse is stated.

function as a result of the relationship between X and Y. In order to retain canonical representations of our algebraic expressions, we must perform some trick such as replacing Y^2 by X^2-1 wherever it occurs. This must be done at a very low level in all parts of the system, otherwise our algebra system will find that the matrix

$$\begin{pmatrix} X-1 & Y \\ Y & X+1 \end{pmatrix}$$

has determinant $(X^2-1)-Y^2 = 0$ but rank 2 because the rank calculations did not apply this equivalence at the right stage. This makes it very difficult to add algebraic relationships to a system which does not include a good treatment of algebraic expressions and the relations to which they give rise. In particular the usual method of introducing square roots into REDUCE-2 (Hearn,1973) by means of a rule such as

FOR ALL X LET SQRT(X)**2=X;

will not work in similar complicated cases, because the transformation is not applied at the correct level.

Replacing high powers of Y (in our example Y^2) absolutely everywhere can cause problems in some cases: consider attempting to take the g.c.d. of $X^2-X*Y+1$ and $X*Y$ by a standard Euclidean algorithm, with X considered to be a more significant variable than Y, i.e. working in $K[Y][X]$. In the obvious notation, the procedure goes as follows:

Explanation	U	V
Initial state	$X**2-Y*X+1$	$X*Y$
$U:=U*Y-V*X$	$-X*Y**2+Y$	$X*Y$
	$=-X**3+X+Y$	$X*Y$
$U:=U*Y+V*X**2$	$X*Y+Y**2$	$X*Y$
	$=X**2+X*Y-1$	$X*Y$
$U:=U*Y-V*X$	$X*Y**2-Y$	$X*Y$
	$=X**3-X-Y$	$X*Y$
Continuing Loop	...	...

More complex cases can be adduced to show that ordering Y before X does not always avoid these problems.   In actual fact, since $X^2-1 = Y^2 = (X-1)*(X+1)$, and Y, $X-1$, $X+1$ are all irreducible, $K[X,Y]$ is not a unique factorisation domain, so that we cannot successfully take g.c.d.s in the domain $\{K[X,Y] \mid F(X,Y)=0\}$. Any representation based on g.c.d.s will fail unless we have a very clear idea of precisely what we mean by a g.c.d., and what algebraic dependencies we are to consider while doing so.

This makes it more difficult to devise a canonical representation for expressions in $\{K(X,Y) \mid F(X,Y)=0\}$.   In my REDUCE-2 programs[*] to perform algebraic-geometric computations I have adopted the following representation, which is based on restricting F to be a polynomial in X and Y, monic in Y. If F is not monic, i.e. it has a leading coefficient G which is not a unit in $K[X]$, then we can write $Y'=G*Y$, and then Y' satisfies a monic equation $F'(X,Y')=0$ and $\{K(X,Y) \mid F(X,Y)=0\}$ is isomorphic to $\{K(X,Y') \mid F'(X,Y')=0\}$.   Furthermore, in most algebra systems (and certainly in REDUCE-2) elements of a field K are represented as the quotient of two elements of an integral domain L (e.g. rationals are the quotient of two integers, or rational functions are the quotient of two polynomials).   We then insist (for purely practical reasons) that all the coefficients of $F(X,Y)$ lie in L (if not, multiply by all the denominators, and then deal with the non-monicity of the equation as described above).   This leads to the assertion that $L[X,Y]^2 \leq L[X,Y]$. The technique of extending by non-monic polynomials is used by Cohen & Yun (1979) to introduce particular fractions into integral domains, but we do not require this, since we always assume that we have a field available to work in.

If these assumptions are satisfied, then an element of $\{K(X,Y) \mid F(X,Y)=0\}$ can be represented as A/B, where B is in $K[X]$ and A is in $\{K[X,Y] \mid F(X,Y)=0\}$ and A and B are coprime as elements of $K[X,Y]$ (i.e. ignoring the relation between X and Y) and B is normalised in some sense (e.g. leading coefficient of B positive) to prevent one multiplying A and B by some element of K to obtain a different representation of the same element of $\{K(X,Y) \mid F(X,Y)=0\}$.

---

[*] Further details can be found in Appendix 1.

Uniqueness of Algebraics

However, being able to manipulate algebraic expressions in the manner described above is not sufficient unless one has a guarantee that they are well-formed (in the sense that $f(X,Y)$, regarded as a polynomial in Y, has no roots in $K(X)$), for otherwise one has no guarantee of canonical representation even if one is using the system described in the previous paragraph. While $F(X,Y) = Y^2 - X^2 + 2X - 1$ is a perfectly reasonable algebraic relationship, and defines an algebraic curve, the curve is not irreducible and the algebraic extension of $K(X)$ defined by that equation is in fact $K(X)$, since the equation factors into linear factors. Similar considerations are brought out in the integration algorithm (Risch,1970), where one has to express certain residues as elements of a Z-module, and therefore has to recognise that $\sqrt{-1}$ and $\sqrt{-4}$ are linearly dependent over the integers[*]. Similarly, the correct algebraic-geometric model for the integration of

$$\sqrt{X^2 - 1} + \frac{1}{\sqrt{1 - 1/X^2}}$$

is not $\{K(X,Y,Z) \mid Y^2 - X^2 + 1 = 0 \ Z^2 - 1 + 1/X^2 = 0\}$ with the integrand being $Y + 1/Z$, but $\{K(X,Y) \mid Y^2 - X^2 + 1\}$ with the integrand being $Y + X/Y$.

In order to test that an algebraic expression is well-formed, we need to ensure that $F(X,Y)$ is an irreducible[$] polynomial (strictly speaking, has only one factor involving Y

---

[*] This is brought out forcibly in Chebyshev's integral (Appendix 2, Example 5), where, having discovered that SQRT(2) is an independent algebraic entity over Q, we then have to consider SQRT(4 SQRT(2) + 9). This is not in fact independent over Q(SQRT(2)), since it is 2 SQRT(2)+1.

[$] This glosses over one point. We need to ensure that it is *absolutely irreducible* (i.e. irreducible over the algebraic closure of the constant field). Now if $F(X,Y)$ is such a polynomial, choose a value y of Y such that $F(X,y)$ is square free (possible if $F(X,Y)$ is). Then if $F(X,Y)$ is absolutely irreducible, it is irreducible over the splitting field of $F(X,y)$, because any factorisation of $F(X,y)$ which corresponds to a factorisation of $F(X,Y)$ can be "grown up" to that factorisation without extending the constant domain (see any p-adic factorisation technique, e.g. Wang (1978) or Yun (1973,1976)).

effectively: but any factor not involving Y might as well be removed anyway). If K is Q (or a purely transcendental extension of Q), then this is the problem of factorising multi-variate polynomials over the integers, which is known to be soluble (Wang & Roths-child,1975 and Yun,1976). If K is not purely transcendental, then we need to use algor-ithms which have recently been developed (see Trager,1976) for factoring over algebraic fields. If we have purely radicals, rather than general algebraic expressions, then we could use many of the algorithms of Zippel (1977).

Representations of Algebraics

If we wish to deal with more than one algebraic expression, we have a choice: we can deal with both of them directly in a *multivariate representation* , or we can express them both in terms of one (more complicated) algebraic expression, viz a primitive[#] element for the field (we will term this a *primitive representation*). The mathematical theory is nearly always cast in terms of primitive elements, since this makes the notation and descriptions easier.   However primitive elements have their disadvantages: with them one would represent $\sqrt{2} + \sqrt{3} + \sqrt{5} + \sqrt{7} + \sqrt{11}$ as a polynomial of degree 31 in a primitive element which satisfied an equation of degree 32. This can prove extremely expensive, and furthermore it must be converted back into the other form if it is going to make sense to a user.   Also it may happen that a problem posed in terms of two algebraic expressions (e.g. SQRT(X**2−1), SQRT(X**2−2)) decomposes into two sub-problems, each of which involves only one of the algebraic expressions.   In a system in which the two expressions are stored separately it is extremely easy to take advantage of this decomposition, while a primitive element system would find that very hard.   For these reasons I use a multivariate expression rather than a primitive element expression, though this decision is largely a matter of opinion. This question was also considered by Caviness & Rothstein (1976), who came to the same conclusion, but also without any hard theoretical evidence. It would be

---

[#] See van der Waerden (1949), Vol I, pp. 126,127 (applying the fact that all fields of characteristic 0 are perfect, and hence have only separable extensions, from pp. 124-125).

interesting to produce such evidence. It is worth noting that the discussion in Chapter 7 accompanying the algorithm DENOMINATOR_ALGEBRAIC, which constructs a defining equation for an algebraic number in order to determine its denominator, is not really evidence for either side, since the same construction would still be necessary, and only marginally easier to program, in a primitive representation system.

If a multivariate representation is chosen, then a general field looks like $\{K(X,Y,Z) \mid F(X,Y) = 0, G(X,Y,Z) = 0\}$, where F involves[#] X and Y effectively, and G involves Z and one of X and Y (but not necessarily both) effectively (there is a discussion of such representations by Shtokhamer(1977)). One example of this is $\{K(X,Y,Z) \mid Y^2-1-X^2 = 0, Z^2-2-X^2 = 0\}$. If we choose this representation, we need a way of deciding whether an algebraic expression is independent of the previous ones, i.e. a mechanism for factoring polynomials over multivariate algebraic fields. This can be obtained from the work of Trager(1976), whose algorithm ALG_FACTOR (see also under Algorithm ALG_FACTOR_2 in Appendix 3) reduces the problem of factoring over $\{K(a) \mid f(a) = 0\}$ to the problem of factoring over K, and this can be applied repeatedly to reduce the problem to factoring over $K(X)$, and this was solved above.

The above remarks might lead one to believe that a system with "good" support for algebraic objects is sufficient for an algebraic-geometric system. That this is not completely true is demonstrated by the following example of the behaviour of algebraic expressions under transformations of the type commonly performed: $\sqrt{x^2-1} \rightarrow$ (under $x \rightarrow 1/x$) $\sqrt{1/x^2-1} = \sqrt{1-x^2}/x = i\sqrt{x^2-1}/x \rightarrow$ (under $x \rightarrow 1/x$ in order to reverse the effect of the previous transformation) $-\sqrt{x^2-1}$, thus proving that $1 = -1$. The solution to this difficulty is to have a separate basis for the algebraic objects in the system for each "value" of x, so that, in the above example, we would not express $\sqrt{1-x^2}$ as $i\sqrt{x^2-1}$ since we have no reason to do so. Thus some form of programmer control over the system's handling of algebraic entities is required. This effect is not wholly destructive, however,

---

[#] Note that we are not allowing F to involve Z at all. We require each algebraic to be determined, in terms of the previous ones (and hence in terms of X) by precisely one equation. If this restriction were to be relaxed, we would need to consider the whole question of Groebner Bases (see the discussion in Buchberger,1979).

since current algorithms for determining dependencies between algebraic expressions (Trager,1976) have complexity at least exponential in the number of algebraic expressions to be considered. Therefore the fact that we need not consider algebraic expressions at one value of x when working at another can lead to an enormous reduction in complexity.

Implementation Considerations

Here we consider the practical implications of the previous paragraphs, and how the processes described are implemented in the program (see also Appendix 1 for more technical discussions of the changes made to REDUCE-2).

The most important remark is that the program deals only with square roots, i.e. an equation $F(X,Y)$ has to be of the form $Y^2 - G(X)$. Since we have a multivariate representation of algebraic expressions, we can represent several square roots simultaneously, and we can represent nested square roots. The decision to impose this restriction was taken on purely practical grounds: all the problems of integration can be illustrated by examples expressed in terms of square roots, and code which is limited to square roots is significantly shorter* and faster than more general code. Since the development of this program has severely strained the computing resources available, this consideration was extremely important.

Corresponding to each "basic-place" (this term is explained in the next section, but can be viewed as a value of X, with corresponding values of $Y,...$), we maintain a separate list of algebraic expressions which determine the field $\{K(X,Y,...) \mid F(X,Y) = 0...\}$, each expression determined by an equation irreducible over the extension defined by all its predecessors. Any time a new algebraic expression is considered, it is tested for independence with respect to the list corresponding to the correct basic-place. This mechanism

---

* The main reason for this is that a square root has precisely one conjugate, which can be obtained by negating the original expression, whereas a general algebraic expression can have several. Furthermore the conjugates of a general expression need not even be defined over the field in which the expression lies.

avoids the difficulty described at the end of the previous section, and ensures that we do not consider algebraics for independence when this is not necessary.

We do not in fact work in terms of variables $X, Y, \ldots$; we work in terms of X only, and Y (where $F(X,Y) = Y^2 - G(X) = 0$) is represented by the expression SQRT(G(X)), which is a REDUCE-2 kernel (Hearn,1973), i.e. it can be regarded as a separate variable for the purpose of forming polynomials and rational functions. The simplification rule

FOR ALL Z LET SQRT(Z)**2=Z;

is declared, and calls (in principle. The actual program has been optimised: see Appendix 1 item 4 for details) to the simplification routine are then written into the code as necessary. This process is unfortunately error-prone, and many of the bugs in the code have been caused by simplifying when I ought not to have done (i.e. working in $\{K(X,Y) \mid F(X,Y) = 0\}$ when I should have been in $K(X,Y)$) or vice versa.

The problem of needing to write simplification calls "sometimes" is one of the greatest problems with the use of REDUCE-2 for algebraic computations, and many other computer algebra systems have the same problems. One of the design aims of the SCRATCHPAD/370 system now under development at IBM Yorktown Heights (Jenks, 1979 and Davenport & Jenks, 1980) is the elimination of this confusion by the provision of a much wider range of possible domains for computation.

Algebraic Geometry

We can view the relationship $F(X,Y) = 0$ as the defining equation of an algebraic curve in the space $K(X,Y)$, and this will prove a very useful point of view in the rest of this work. For a view of Algebraic Curves and Algebraic Geometry at the level we will need, see (Fulton,1969) or (Seidenberg,1968). A completely algebraic approach can be found in (Chevalley,1951), while a hybrid approach is contained in the book by Eichler (1966). The algebraic curve could be viewed as a Riemann Surface with branches, but in fact the standard terminology of algebraic geometers is to talk about "places", a place being the same as a branch, although the standard definitions are in terms of local valuation rings.

We speak of a place *lying over* a value A of X (possibly including infinity), or being *centred at* that value, to mean that the function $X-A$ (or $1/X$ in the case of infinity) takes the value 0 at that place.

Such an algebraic curve can have multiple points, but corresponding to every algebraic curve (and to every finitely generated algebraic extension of $K(X)$) there is a "non-singular model" of the curve (see Fulton,1969 Chapter 7, Theorem 3 & Corollary), which is obtained from the original curve by "blowing up" all the singularities on it. This picturesque terminology means that the point in the plane which was the multiple point has been replaced by a line, and the multiple point on the curve has been replaced by several simpler points, and if this process is repeated sufficiently often, we will end up with only simple points. Unfortunately this non-singular model is no longer a curve in the X,Y plane, and it is hard to visualise or manipulate this curve directly. So this Theorem about the existence of non-singular models is in fact not of great computational use.

Geometrically, there are two possible types of behaviour at a multiple point, depending on whether or not the tangents are multiple. A point of multiplicity 2 with 2 separate tangents can be seen at the origin in $Y^2 = X^2 - X^3$, whereas $Y^2 = X^3$ has a point of multiplicity 2 with one tangent at the origin. Of course it is possible for both types of behaviour to exist simultaneously at the same point, (e.g. a point of multiplicity 7 with 3 tangents), but this is not inherently more complicated. A point of multiplicity n with n distinct tangents to the curve at it is termed an *ordinary* multiple point. A much more useful transformation than the one described in the previous paragraph is one that transforms a plane curve to a birationally equivalent plane curve, all of whose multiple points are ordinary (see Fulton, 1969, chapter 7). In practice, it is normally sufficient to do this type of de-singularisation "locally", i.e. to transform a non-ordinary multiple point away if one finds that a particular multiple point is troublesome, rather than attempt to find all the multiple points and perform this transformation at all of them. These techniques, plus a recognition of ordinary multiple points, allow one to perform calculations on algebraic curves without being bedevilled by the potential singularities, although the code must always allow for the possibility of there being a singularity and make the necessary transformations. One other kind of transformation which we may frequently wish to make is the *birational transformation*, which is a map from a curve in $(x_1,...,x_n)$ space to a curve in $(y_1,...,y_m)$ space such that all x_i are rational functions of the y_j and vice versa. If there is a birational transformation from A to B, we say that these are *birationally equivalent*.

Lemma 1 * Two curves are birationally equivalent iff their function fields are isomorphic.

Puiseux Expansions

A major tool in Algebraic Geometry is the theory of *Puiseux Expansions*. A Puiseux Expansion about a place is the equivalent of a Laurent Expansion about a point in ordinary function theory. Just as in conventional function theory we cannot express $\sqrt{X}$ as a Laurent Series (with integral exponents) in terms of X, so on $\{K(X,Y) \mid Y^2-X = 0\}$ we cannot express Y as a Puiseux expansion in terms of X about $X=0$, even though we can express X in terms of Y about the same point. However, for any place centred at A, there is always some fractional power of $(X-A)$ (or $1/X$ in the case on infinity) such that all functions in $\{K(X,Y) \mid F(X,Y) = 0\}$ can be expressed as Puiseux expansions in terms of it (see Chevalley, 1951, Chapter I, Theorem 2). Any function such that all functions can be expressed as Puiseux expansions in terms of it about a place, but such that no integral power of it will suffice, is termed a *uniformising variable* or *local parameter*. We shall frequently need to bear in mind the fact that $(X-A)$ or $1/X$ need not be a local parameter. A place at which it is not a local parameter is said to be *ramified*, and the power of the local parameter which gives $(X-A)$ or $1/X$ is said to be the *ramification index*[$] (and hence an unramified place has ramification index 1).

We define the *order* of a function at a place to be the index of the first non-zero term

---

* Fulton (1969) Proposition 12, p. 155.

[$] The terms "ramified" and "ramification index" are also used in algebraic number theory with somewhat different meanings, and we shall need these other meanings (which we shall define there) in Chapter 8. It should always be clear from the context whether we are using the geometric or number-theoretic meanings.

in its Puiseux expansion about this place divided by the ramification index of that local parameter. This is clearly independent of the local parameter chosen.

A Data Structure for Places

We require some form of data structure to represent the mathematical concept of a place in our program. The overall format of the data structure I have chosen is that of a LISP substitution list, i.e. a list of pairs (old value . new value).

The first pair in the list determines the value of the variable of integration (which we shall assume to be X for the purpose of all the examples) over which the place lies. If the place lies over $X = 0$ the first pair is (X.X), if the place lies over the point at infinity, the first pair is (X QUOTIENT 1 X), and if it lies over the point $X = a$, the first pair is (X PLUS X $-$ a). After this substitution has been performed, the variable X now takes the value 0 at the point over which the place lies.

However X need not be a local parameter. The second element of the list is a substitution pair designed to rectify this problem (if it exists) and is (X EXPT X n), where $X^{1/n}$ is a local parameter at the place in question. If X was already a local parameter, this component is not present. This part (if present) and the previous part constitute a *basic-place*, i.e. an expression for a local parameter at a point such that all that is required to convert it into a description of a place is some way of determining which place over that point is intended.

The remainder of the list is designed to discriminate between the various places lying over one point. This can be much more difficult than it seems, for the places are in some sense indistinguishable, all being roots of the algebraic equation defining the curve. A purely arbitrary decision can be made, calling one of the roots[*] of $Y^2 = X^2 - 1$, for example, SQRT(X**2−1), and the other one −SQRT(X**2−1). Then this part of the list

---

[*] In the case where $F(X,Y)$ is not of the form $Y^2 - G(X)$, but is of degree n in Y, we will have n possibilities, corresponding to the conjugates of Y.

contains substitution pairs of the form (Y.Y) or (Y MINUS Y), where Y is an expression of the form SQRT(polynomial in X). Y is expressed as if the transformations in the first and second part of the list had been conducted, e.g. the two places of $Y^2 = X^4 + X^3 + 1$ lying over infinity are ((X QUOTIENT 1 X) (expr.expr)) and ((X QUOTIENT 1 X) (expr MINUS expr)), where expr is SQRT($1+X+X**4$).

With this data structure, the evaluation of functions at places ought to be straightforward, in that one should substitute $X=0$ in the result of substituting the expression for the place into the function. Unfortunately, this frequently gives 0/0 as the answer.

The immediate answer to this is to use l'Hopital's rule for the evaluation of indeterminate limits, viz. that $\lim F(X)/G(X) = \lim F'(X)/G'(X)$ if $G(0)=0$. While this rule is extremely useful to the mathematician, there is no guarantee that $F'(X)/G'(X)$ is easier to evaluate than $F(X)/G(X)$, and it may well be harder - the reader may care to consider

$$\frac{iX-\sqrt{2}\sqrt{A-\sqrt{A^2+X^2}}}{A+\sqrt{X^2+A^2}}$$

where F' and G' are both more complex than F and G. The answer to this problem is in fact to calculate such values as the coefficient of X^0 in Puiseux expansions. Harrington, in his work (Harrington, 1979a) on the evaluation of limits, has a variety of applications of l'Hopital's rule, but ends up admitting that Taylor series (equivalent to Puiseux expansions) may be necessary for the evaluation of some limits.

Computation of Puiseux Expansions

The computation of Puiseux expansions is a problem very similar to the computation of Taylor (or, more strictly, Laurent) Series. Before this problem is dismissed as totally trivial, it is worth considering the question "What is the coefficient of X^0 in

$(1+Y)/(1-Y)$, where $Y^2 = X^2 + 1?$" #. A straightforward computation will give us $2/0$, and in order to obtain the correct answer we have to discover the coefficient of X^2 in $1+Y$ and $1-Y$. Also, having computed a power series as far as the X^4 term, say, we may need the X^5 term, and it would be wasteful to recompute the entire series in order to obtain the next term.

The best technique for expanding power series in this way is the "incomplete but exact" approach of Norman (1975). Norman summarises the design by saying that his package "gives the impression that it is computing with full rather than truncated power series. Calculations are only performed when an attempt is made to display or use the results, and so no unnecessary work is ever done".

Norman was able to use the SCRATCHPAD system of rules and values in order to produce an extremely neat implementation. We do not have this mechanism available in REDUCE-2, but we can perform equivalent functions. The representation of a Puiseux Expansion is a data structure with two components: a list of already computed terms, and a method of evaluating terms in it, which will be an operator $(+,-,*,/,SQRT)$ and a list of arguments, which are themselves Puiseux Expansions of components of the original expression.

Divisors

We can construct the free Abelian group on all the places of an algebraic curve, i.e. the set of all finite collections of places with integer multiplicities. An element of this group is termed a *divisor*. This group can be written either multiplicatively or additively, and the literature has both. We will normally find it convenient to write it multiplicatively. The sum of all the multiplicities is termed the *degree* of the divisor. One way of regarding the multiplicities in a divisor is to say that they tell you whether this place is a zero (of the multiplicity specified in the divisor) or a pole (with the multiplicity specified in the divisor,

---

The answer is in fact -2.

out ignoring the − sign) of a specified function. In this case the divisor D is termed the *divisor of the function f* specified, and we write $D=(f)$. Every function except the zero function has a divisor, and the mapping from the multiplicative group of non-zero functions into the group of divisors is in fact a group homomorphism. The kernel of this homomorphism is precisely the set of constant functions (i.e. the non-zero elements of K). The divisor of a function always has degree 0 (Chevalley, 1951 Chapter 1, Theorem 5), but the converse is not always true. In the special case of $K(X)$ it is true, for one multiplies together the $(X-A)^N$ for all finite places lying over A with multiplicity N, and this necessarily has the correct multiplicity over infinity. A divisor with no negative multiplicities is said to be *effective*.

In general there are three possibilities for a divisor of degree 0: it can be the divisor of a function in $\{K(X,Y) \mid F(X,Y) = 0\}$, when we say that the divisor is *principal* or *linearly equivalent* to zero; there can be a power of it which is linearly equivalent to 0, when we say that the divisor is *rationally equivalent* to 0, or a *torsion divisor* ; or there can be no power of it which is the divisor of a function, when we say that the divisor is (rationally) *inequivalent* to 0. If the divisor is rationally equivalent to 0, then the power to which it must be raised to make it linearly equivalent to 0 is termed the *order* of the divisor.

In the case of $K(X)$, not only are all divisors of degree 0 linearly equivalent to 0, but, as shown above, it is trivial to find the function which corresponds to a divisor of degree 0. In general this is not true, and the breakthrough in finding such a function, if the divisor is linearly equivalent to 0, was the work of Coates (1970), described in the next chapter.

Differentials

Another useful concept in Algebraic Geometry is that of the *differential*, which can be thought of as an expression of the form $f(X)dX$, even though a rigorous definition has to be much more abstract (see Fulton,1969, pp. 203-5). We regard differentials as being the same if the ordinary rules of schoolboy calculus would make them so (see Fulton,1969 again for a precise description), e.g. $d(X^2) = 2XdX$. We can make the differentials into a

$\{K(X,Y) \mid F(X,Y) = 0\}$ module in the obvious way, e.g. $c\,\mathrm{d}X = \mathrm{d}(cX)$ for c in K. The space of all differentials is a 1-dimensional vector space over $\{K(X,Y) \mid F(X,Y) = 0\}$, and $\mathrm{d}X$ is a basis for it.

If t is a uniformising parameter about a place P, then any differential can be written as $f\mathrm{d}t$ for some f in $\{K(X,Y) \mid F(X,Y) = 0\}$. Define the *order* of the differential at P to be the order of f at P (this is well-defined by Fulton, 1969 Chapter 8, Prop. 7). Hence we can define the *divisor of a differential* in the same way as we defined the divisor of a function earlier. Note that the two are different ideas: the divisor of $1/X$ is a pole of order 1 at 0 and a zero of order 1 at infinity, while the divisor of $1/X\,\mathrm{d}X$ is a pole of order 1 at 0 and a pole of order 1 at infinity (because $t = 1/X$ is a uniformising parameter there, and $1/X\mathrm{d}X = t(\mathrm{d}X/\mathrm{d}t)\mathrm{d}t = -1/t\mathrm{d}t$).

A differential is said to be *of the first kind* if it has no poles. An example of this is $1/Y\,\mathrm{d}X$ when $F(X,Y) = Y^2 - (X^3 + 1)$. Differentials of the first kind form a vector space over K, and the dimension of this space is termed the *genus* or *deficiency* of the curve, often denoted by g. The genus is a parameter of the curve of great importance in Algebraic Geometry. In particular, a curve of genus 0 can be transformed into a line by means of a *rationalising substitution*. This is the justification for the use of trigonometric substitutions to integrate functions defined over curves of genus 0 (of which $F(X,Y) = Y^2 - (1 - X^2)$ is the simplest example), because if X is cos t, then Y is sin t, and both of these are rational functions of tan $t/2$, which is an actual rationalising parameter. A curve of genus 1 is said to be an *elliptic curve*. We will discuss these curves in greater detail in Chapters 5 and 7.

As in classical function theory, we can define the *residue* of a differential at a place. There are a variety of abstract definitions, but for our purposes the definition which says that it is the coefficient of t^{-1} in an expansion of the differential in terms of a uniformising parameter t is sufficient. If the residue of a differential is non-zero then that place must be a pole of the differential, so a differential can only have a finite number of non-zero residues. A differential, all of whose residues are zero, is said to be *of the second kind*. Clearly every differential of the first kind is of the second kind also. The differentials of he second kind form a $2g$ dimensional vector space over the space of exact differentials. A differential is said to be *of the third kind* if all its poles have order 1. Differentials which are of both the second and the third kind must be of the first kind.

The representation of a differential is precisely the same as that of a function: however the rules for transforming and evaluating it are not the same. Because a differen-ial is $f\,dt$ for some function f and uniformising parameter t, when we wish to evaluate the differential at a place we must transform the function f (as described in the section "A Data Structure for Places" above), and also multiply by dt/dt', where t' is a uniformising parameter at the new place.

Coates' Algorithm

Introduction

In this chapter, we consider the problem of finding a function with a certain set of poles. That this problem is non-trivial in the case of algebraic functions (although it is trivial in the case of rational functions) can be seen from the fact that such functions need not always exist. For example, on the curve defined by $\sqrt{X^3 + 1}$, there is no function with a zero of order 1 at one place lying over the point $X=0$ and a pole of order 1 at infinity and no other poles or zeros, but there is one with divisor 3 times that (i.e. the divisor has order 3). On the curve defined by $Y^2 = X^3 - 3X^2 + X + 1$, there are no functions with a zero at one place lying over $X=0$ and a pole at the other, both having the same order, and no other zeros or poles (see Appendix 2 example 7).

Coates published[*] an algorithm in 1970 (Coates,1970) to find functions on algebraic curves over algebraic number fields as part of a bigger project on Diophantine equations (Baker & Coates,1970), and this context coloured his precise formulation of the results he stated and proved. We will wish to use a slightly different statement of the algorithm in a slightly different context, as described below. In particular, in view of the remarks in the last chapter, we can assume that no place at which we wish to compute is ramified, although the theory only requires this to be true of infinity (and, even if this is not so, Coates (1970) Theorem 2 implies that a slightly modified result is still true). For computational convenience, we will often use an unramified representation even when the theory does not require it.

---

[*] Coates only stated his algorithm for algebraic number fields, and was only interested in showing that it was mathematically effective. The extension to arbitrary ground fields is, I believe, original (though not profound). I also believe that this work describes the first computer implementation of Coates' Algorithm.

The Algorithm

Let D be a divisor of poles (i.e. all multiplicities in it are negative) all of whose places are finite. Then the algorithm will produce a basis for the K vector space of all functions f with $(f) \geq D$, i.e. the space of functions with poles no worse than D. This algorithm relies on two subsidiary ones, INTEGRAL_BASIS_REDUCTION and NORMAL_BASIS_REDUCTION, which, following the dictates of the top-down programmers, are described after the main algorithm. These algorithms are described in simplified form, in that we do not mention the complications caused by the necessity to take representations of our curve on which certain places are unramified. We also do not describe several of the techniques used for efficiency, since they complicate the algorithms unnecessarily for our purposes, even though they are important from a practical point of view.

The algorithm can be summarised as follows: we start with any list of functions which has no finite poles, progressively insert all the finite poles as specified by D (using INTEGRAL_BASIS_REDUCTION), and then find those resulting functions that have no poles over infinity (using NORMAL_BASIS_REDUCTION).

Before we can define this algorithm fully, we need a couple of definitions. Let V be a module over the (commutative) ring R. We say that $(v_1,...,v_k)$ form a *basis* for V over R if all the v_i lie in V and each v in V can be expressed in the form $rv = \sum_{i=1}^{k} r_i v_i$ where r and the r_i are in R, with r non-zero, and all such expressions are unique up to multiplication by elements of R. A basis is said to be an *integral basis* iff r can be chosen to be 1 in all such expressions, i.e. each element of V can be expressed as a linear combination of the v_i with coefficients in R.

COATES

Input:

F(X,Y): the equation of the curve

(possibly in a multivariate representation, e.g. $F(X,Y) = 0$, $G(X,Y,Z) = 0$).

D   : a divisor of finite poles on the curve

Output:

V : A basis for the K-space of functions f: $(f) \geq D$

[1] V:=(1,Y,Y**2 ... Y**(degree of F in Y −1))

In the event of a multivariate representation, we let V be the set of all possible products of the algebraic variables raised to powers less than their degree.

V is now a basis (but not an integral basis - see above for a precise definition of this term) for the $K[X]$ module of functions with no finite poles worse than D (or indeed worse than 0).

[2] For Each P over which places in D lie

V:=INTEGRAL__BASIS__REDUCTION(V, P, that part of D which lies over P).

Each step of this makes V have the right multiplicities at those places of D which lie over P. V is now a basis for the $K[X]$ module of functions with no finite poles worse than those in D and which is "quasi-integral" in the following sense: any member of that module can be expressed as a linear combination of elements in V with coefficients from $K[X]$, possibly divided by a member of $K[X]$, but this member of $K[X]$ will have no zeros at the (finite) places which occur in D.

[3] Now ensure that this really is an integral basis.

[3.1] Z:=An arbitrary element of K (e.g. a "GENSYM").

[3.2] A:=determinant A(I,J), where

A(I,J) = coefficient X**0 in V(I) expanded about J'th place over Z.

3.3] For all roots (in Z) of the numerator or denominator of A

        V:=INTEGRAL__BASIS__REDUCTION(V, Z, 0 divisor over Z).

4] V:=NORMAL__BASIS__REDUCTION (V,places of curve lying over infinity).

This ensures that, not only do we have an integral basis, but that its elements are of as high an order (i.e. have as little of a pole) as possible at infinity. The elements of this basis V, when multiplied by suitable powers of X, are the required basis

5] V:=For Each V(I) with order $N \geq 0$ at infinity

        Collect (V(I),V(I)*X,.....,V(I)*X**N).

In fact N must be the minimum order of V(I), over all places lying over infinity.

I have currently no computing time analysis for this algorithm, but in cases arising in practice, step 3 is nearly always the most expensive operation. This step is needed to ensure that we have not inadvertently started with, or introduced later on, a dependence between the elements of V over some point Z which did not appear in the divisor D. Under many special formulations of the problem one can prove that this cannot happen, and I conjecture that there is some wide class of problems for which this is true, and that it should be possible to reduce all problems to one of this kind. However, I have not made much progress in this area.

If one excepts step 3, then it might appear that the time would be proportional to the sum of the multiplicities in the divisor, since this is the number of individual INTEGRAL__BASIS__REDUCTION steps that are performed, to within some complicated edge effects. Things are not quite so simple, however, because such steps tend to increase the size of the elements of V. I believe that the time is, in fact, exponential in the sum of the multiplicities, but the underlying analysis is not rigorous. The dependence of the time on the degree of $F(X,Y)$ is clearly much worse, but here again I have no definite formula. One of the unfortunate features of this algorithm is that one starts with a fairly simple basis in step 1, step 2 produces an integral basis with enormous expressions, and step 4 then reduces them drastically in size, often to the same complexity as the original basis. Thus it would appear that there is a much better formulation of this fundamental algorithm waiting to be discovered.   Sometimes performing reduction steps at infinity (which can be done in the middle of reduction steps at other points) can control this growth, but frequently it merely wastes time (since the reduction at infinity must be performed at the end as well) and does not reduce the expression size at all.

INTEGRAL__BASIS__REDUCTION

Input:

V: a basis for the $K[X]$ module.

P: a value of X to ensure that the basis is integral at.

D: a divisor of poles lying wholly over P.

Output

V: a modified basis, integral over P.

[1] Let A(I,J) be the coefficient of $(X-P)**N(J)$ in the expansion of V(I) about the J'th place in D, where N(J) is the multiplicity with which that place occurs.

There might be problems if $(X-P)$ is not a local parameter, so in fact we may need to consider some fractional power of $(X-P)$.   However, the algorithm is easier to describe without this complication.

[2] While A is singular do:

[2.1] Let B(I) be elements of K such that SUM(B(I)*A(I,J))=0 for all J, but the B(I) are not all zero.   In particular, assume B(K) is non-zero.

[2.2] V(K):=SUM(B(I)*V(I))/(X−P)

Then the V(I) are still linearly independent over $K(X)$ and have poles no worse than D at P.

[2.3] Recompute A(K,J) as in [1].

[3] Return V

NORMAL_BASIS_REDUCTION

Input:

V: an integral basis for the $K[X]$ module

D: The places of the curve lying over infinity.

Output

V: A normal integral basis for the K[X] module

[1] Sort the V(I) in decreasing order of their minimum order at all of the places over infinity.

[2] Let A(I,J) be the coefficient of $1/X**0$ in the expansion of V(I) about the J'th place in D.

There might be problems if $1/X$ is not a local parameter, so in fact we may need to consider some fractional power of $1/X$. However, the algorithm is easier to describe without this complication.

[3] While A is singular do:

[3.1] Let B(I) be elements of K such that $SUM(B(I)*A(I,J))=0$ for all J, but the B(I) are not all zero. In particular, assume B(K) is non-zero.

Among all such possible B(I), we want the one involving the V(I) of highest order, i.e. the first one to be discovered if one tries Gaussian elimination without interchange on A.

[3.2] $V(K):=SUM(B(I)*V(I))/X$

Then the V(I) are still linearly independent over $K(X)$ and have poles no worse than D at P.

[3.3] Sort the V(I) as in [1].

[3.4] Recompute A(K,J) as in [2].

[4] Return V.

Proof of Algorithm [1] - [3]

Let us first introduce some notation. Assume K is an algebraic field, and let L be its algebraic closure. Let $F(x,y)$ be the defining equation of our algebraic curve* whose coefficients lie in K, and let it have degree $n \geq 1$ in y. We assume that F is irreducible over L. Furthermore, let it have n distinct unramified places Q_i lying over infinity. Let R be the field $\{L(x,y) \mid F(x,y) = 0\}$. Suppose we are given $s \geq 1$ distinct elements of L, a_h ($1 \leq h \leq s$), and we denote by $A_{hi} (1 \leq i \leq r_h)$ the places+ of R lying over the (finite) point $x = a_h$ and we denote their ramification indices by e_{hi} respectively, so that $\sum_{i=1}^{i=r_h} e_{hi} = n$. Finally we assume that we are given positive integers v_{hi} for each A_{hi}, and we let M be the set of all g whose order at each A_{hi} is $\geq -v_{hi}$ and whose order at each other place of R is ≥ 0 (including infinite places). Hence we are regarding the A_{hi} and the v_{hi} as a divisor, and M as the set of functions with divisor greater than this divisor of poles.

If a is any member of L, and A_i ($1 \leq i \leq r$) are the places of R lying over the point $x=a$, (with ramification indices e_i respectively) we associate with the A_i ($1 \leq i \leq r$) integers v_i ($1 \leq i \leq r$) as follows: If a is one of the a_h mentioned in the previous paragraph, then $v_i = v_{hi}$, otherwise we let v_i be 0. We will write $\sum_{i=1}^{r} v_i = V$. Now let M' be the set of all g in R whose order at A_i is $\geq -v_i$ for all elements a of L. Thus M' is an $L[x]$ module of dimension n. The first step in the construction of a basis for M over L will be the construction of an integral basis of M' over $L[x]$, where elements $w_1,...,w_n$ of M' are said to form an *integral basis* of M' over $L[x]$ iff they are linearly independent over $L[x]$ and every element of M' can be expressed an a linear combination of them with coefficients from $L[x]$.

Let $w_1,...,w_n$ lie in M'. We can write the expansion of w_j at A_i as $w_j = X^{-v} \sum w_{jik} x_k$

---

* This does not preclude the use of multivariate representations for computational purposes; we merely choose to state and prove these results about Coates' Algorithm in a primitive representation. My code uses multivariate representations throughout its work on Coates' Algorithm.

+ Which Coates refers to as *valuations* in his paper (1970). As far as we are concerned, the two words are interchangeable.

where $X^e = (x-a)$ and X is therefore a local parameter at A with ramification index e. Let $D(a)$ be the determinant whose j-th row consists of the elements w_{jik} with $1 \leq i \leq r$ and $0 \leq k < e_r$ arranged in lexicographic order of indices.

Lemma 1 (this is equivalent to Lemma 6 of Coates(1970)) $w_1,...,w_n$ form an integral basis of M' iff $D(a)$ is non-zero for all a in L.

Proof: To prove the necessity, suppose that $D(a)=0$ for some a in L. Then there exist elements $x_1,...,x_n$ of L, not all zero, such that $\sum\limits_{j=0}^{n} w_{jik}x_j = 0$ for $1 \leq i \leq r$ and $1 \leq k < e_i$, since the rows of the matrix must be linearly dependent over L, but then $w = (x-a)^{-1}\sum\limits_{j=1}^{j=n} x_j w_j$ belongs to M', and hence such that the original w_j were not an integral basis.

Now let us assume that $D(a)$ is always non-zero, and then prove the sufficiency of this condition. Let $d(w_1,...,w_n)$ be the determinant whose ij-th element is $w_i^{(j)}$, where this stands for one of the n field conjugates of w_i belonging to an extension of $L[x]$ in which $F(x,y)$ splits into linear factors and the conjugates are ordered suitably*. Then $d(w_1,...,w_n)$ is an element of $L(X)$ and is not identically zero, for if we choose a so as to have n unramified places above it, we have that $d(w_1,...,w_n) = D(a)(x-a)^V +$ higher powers of $(x-a)$ (since the expansion of w_i at A_j is the formal power series in $(x-a)$ representing $w_i^{(j)}$) and $D(a)$ is non-zero. Since this is so, the w_i form a basis for M', and any w in M' can be written uniquely as $wq(x) = \sum\limits_{j=1}^{j=1} q_j(x)w_j$, where $q(x)$ and the $q_j(x)$ are polynomials in x with no common factor. If we can show that $q(x)$ is a constant, then we will have expressed w integrally in terms of the w_j and the lemma will be proved. So we suppose, if possible, that $q(x)$ has a factor $(x-b)$. Then write $q(x) = (x-b)q'(x)$ and consider

---

* In fact, d is only determined to within a factor of ± 1, since the determinant can be permuted, but this will not worry us. Coates (1970, Lemma 6) defines d to be the square of the determinant in order to avoid this ambiguity, but we will not do so. I am grateful to Dr. Norman for pointing out this simplification.

$$z = \frac{\sum\limits_{i=1}^{i=r} q_i(b)w_i}{x-b} = q'(x) - \frac{\sum\limits_{i=0}^{i=r} q_i(x) - q_i(b)}{x-b}$$

Now each term of the last expression belongs to M', so z belongs to M'. But the $q_i(b)$ are constants, not all zero (for else $(x-b)$ would be a common factor of all the q's, contrary to hypothesis) and $D(b)$ is non-zero, so there is a combination (i,k) with $k < e_i$ such that $\sum\limits_{j=0}^{j=n} q_j(b)w_{jik}$ is non-zero. Therefore z cannot lie in M' - the required contradiction.

This Lemma suggests the idea of a *reduction step*, which consists of replacing one of the w_i in the basis by the w computed in the first portion of the Lemma (equation 1), thus giving a new basis $w'_1,...,w'_n$ which is "closer" to being an integral basis than the initial basis. The algorithm INTEGRAL__BASIS__REDUCTION described above performs as many of these reduction steps as are needed over the point P, having chosen, for computational convenience, a model of the curve over which P is unramified.

Lemma 2 The process terminates after a finite number of reduction steps.

Proof: We first observe that, after a reduction step at a, $d(w_1,...,w_n) = d(w'_1,...,w'_n)(x-a)$ to within a constant factor (i.e. an element of L) since we have replaced one of the w_i by an element with an extra factor of $(x-a)$ in the denominator. Now let $A(x) = \prod\limits_{h=1}^{h=s} (x-a_h)^{v_{hi}}$ Then the elements of $A(x)M'$ have no finite poles at all, and in fact $d(w_1,...,w_n)A(x)$ lies in $L[x]$, since each element in the sum which is the determinant contains one item from each conjugate, and therefore can only contain one contribution of each v_{hi} for each h. Then, since d is decreasing with each reduction step and is bounded below, the process must terminate.

Proof of Algorithm [4] - [5]

We have now constructed an integral basis for the space M', say $(w_1,...,w_n)$. Let the expansion of w_j at Q_i (which are the places over infinity) be given by $w_j = X^{l_j}\sum\limits_{k=0}^{\infty} w_{ijk}X^k$

where $X = 1/x$ is a local parameter at Q_i since it is unramified and l_j is an integer such that w_{ij0} is non-zero for some place Q_i (although not necessarily for all such). The basis is defined to be a *normal integral basis* if the determinant D, whose elements are defined to be w_{ij0}, is non-zero. We will assume that the w_i have been ordered such that $l_i \geq l_{i+1}$ ($1 \leq i < n$) and that the l_i are non-negative for $i \leq l$ (it may be that $l = 0$, of course).

This suggests the idea of a *reduction step at infinity*, in which, if $D = 0$ so that there are $x_1, \ldots, x_n$ with $\sum_{j=1}^{j=n} x_j w_{ij0} = 0$ ($1 \leq i \leq n$), and we let x_h be the last non-zero x_i, we replace the basis $(w_1, \ldots, w_n)$ by a new basis in which w_h is replaced by $w'_h = \sum_{i=1}^{n} x_j x^{l_j - l_h} w_j$ (the exponent of x is non-negative by virtue of the ordering of the w_i).

Lemma 3 If the w_i form an integral basis of M', applying a reduction step at infinity leaves them an integral basis.

Proof: It is clear that the new set of w_i lie in M', so the only question is whether or not they form an integral basis. But a reduction step at infinity does not alter $d(w_1, \ldots, w_n)$ (except by a constant multiple), since w_h is replaced by a linear combination of rows, in which w_h appears with constant coefficient.

Lemma 4 We can only perform a finite number of reduction steps at infinity.

Proof: After a reduction step at infinity, l_h has increased by at least 1, and the rest of the l_i are unchanged (since the corresponding w_i are). Therefore $\sum_{i=1}^{i=n} l_i$ increases by at least 1 on each reduction step over infinity. Conversely $\sum_{i=1}^{i=n} l_i$ is at most equal to the order of $d(w_1, \ldots, w_n)$ at infinity (since $w_j^{(i)}$ has order at least l_j at any of the Q_k) and, as remarked in the previous Lemma, $d(w_1, \ldots, w_n)$ is unchanged after a reduction step, so $\sum_{i=1}^{i=n} l_i$ is bounded above.

Lemma 5 (this is equivalent to Lemma 9 of Coates (1970)) If w_j ($1 \leq j \leq n$) is a normal integral basis of M', then $\{x^h w_j \quad 1 \leq j \leq l \quad 0 \leq h \leq l_j\}$ is an L-basis of M.

Proof: Clearly the elements of this set are linearly independent over L, and lie in M, so all that remains to be proved is that, for any g in M, g can be expressed as an L-linear combination of these elements. Since g is in M', we can write that $g = \sum_{j=1}^{j=n} q_j(x)w_j$ where $q_j(x)$ lies in $L[x]$, since the w_j $(1 \leq j \leq n)$ form an integral basis of M'. Let m_j be the degree of q_j as a polynomial in x, and m be $\min(l_j - m_j)$ over all j with $q_j(x)$ not identically zero. Then we can write $q_j(x)$ as $c_j(1/x)^{m-l_j} +$ higher powers of $(1/x)$. Hence, at Q_i, g is $(1/x)^m \sum_{j=1}^{j=n} c_j w_{ij0} +$ higher terms in $1/x$. Since D is non-zero, at least one of these $\sum c_j w_{ij0}$ is non-zero. Then the order of g at this Q_i is m. Since g belongs to M, we can deduce that $m \geq 0$, and hence that $m_j \leq l_j$ which proves that g lies in the space generated by the set.

Extensions

The version of Coates' Algorithm described above only works for divisors lying wholly over finite points. This is clearly not a fundamental restriction, and many ways could be devised to circumvent it. The most obvious way is to transform the curve, and the space in which it lies, so that the line (or hyper-plane, in general) at infinity no longer passes through any of the places of the divisor. The trouble with this is two-fold: such a transformation is quite difficult when we have more than one algebraic quantity; and it introduces a wholly fictitious set of algebraic entities. Even in view of the remark in the previous chapter under "Algebraic Necessities", which says that these can (and should) be kept distinct from the previous ones, we will still have some work to do to establish irreducibility and to compute Puiseux expansions.

The technique now adopted in the program avoids both these difficulties. Let N be the greatest (i.e. most negative) order at infinity. Then add N to all the orders at infinity (making allowances for any ramification that may be present) and subtract N from all those lying over 0^*. All our poles now lie over finite places, and we can apply Coates' Algorithm in the form stated above. We then multiply all the answers by X^N, and have a basis as required, except that it will be for functions with poles of order N over all infinite places. Reducing these orders to the ones we want is merely a matter of applying some linear constraints, so that a basis for the functions with at worst the desired poles is easily achieved.

We have only considered divisors of poles so far. However, a major use of Coates' Algorithm is to determine whether or not a divisor is the divisor of a function, and for this we must find a basis (which will necessarily be 1-dimensional if it exists) for the functions with the zeros and poles of the divisor. The obvious way, and the one I used to adopt, is to take the basis of functions with at worst the poles from the divisor (as found by the algorithm COATES) and then to regard the zeros as a series of linear constraints#, and to solve the resulting set of linear equations. One might think that this was not guaranteed to work, because not all the functions in the basis produced by Coates' Algorithm have all the poles, so why should that linear combination of them which has the zeros have all the poles? In fact, if the divisor were of degree 0 (and otherwise it cannot be the divisor of a function), then any function produced which has all the zeros must also have all the poles, because it must have that number of poles and it can have no others, since no element in the basis has any others.

This solution works very well in most cases, but there are difficulties of a practical nature which can arise. Consider the problem in $K(X)$ of finding the function with a zero

---

* It is possible that there may not have been any places lying over $X=0$ in the divisor, in which case we have to insert (with multiplicity 0) all the places on the curve lying over $X=0$. It might be more efficient to choose a value other than $X=0$, to avoid having to do reduction over an extra value of X, but I have not done this.

These constraints are of the same type as those which were introduced in the previous paragraph to deal with the movement of poles from infinity, and so we only need to solve one set of linear equations, rather than two.

of order 1000 at $X-1$ and a pole of order 1000 at X (which is clearly $\left(\frac{X-1}{X}\right)^{1000}$). Our basis for the functions with a pole of order ≤ 1000 at X, and no poles elsewhere, could be $1, 1/X, 1/X^2, ..., 1/X^{1000}$. We then have a 1000 by 1000 set of linear equations to solve in order to find the correct function. Even though the derivation of these equations suggests that they must have a great deal of structure, straightforward calculation makes them dense (because the solution is the row vector whose i-th element is the binomial coefficient $_{1000}C_i$, and which has no non-zero entries) and therefore impractical to solve. This particular difficulty could be solved by adapting Coates' Algorithm to deal with zeros as well as poles (a trivial task) and starting off with the function $(X-1)^{1000}$, but this solution does not work for general algebraic problems. Consider the curve defined by $F(X,Y) = Y^2 - (X^2 + 1)$, and try to find the function with a pole of order 1000 at one place over X, and a zero of order 1000 at the other (the answer is $\left(\frac{X}{Y-1}\right)^{1000}$ as the reader may verify). Here our initial guess would have to be X^{1000} and we would then have to perform 2000 reduction steps over X and the intermediate basis would contain dense polynomials of order 2000 in X. These examples with large numbers are given to illustrate that this algorithm and the work based on it suffer very badly from exponential growth.

The following alternative technique[#] (which has now been implemented in the latest version of the integration system) removes all these difficulties. Suppose we wish to find a function with poles at $P_1, ..., P_n$ and zeros at $Q_1, ..., Q_n$. Let f be a function with poles at P_1 and P_2 and a zero at Q_1. Such a function[*] can readily be found by Coates' algorithm. Then let P' be the other zero of f, and f' be the function (found recursively) with poles at $P', P_3, ..., P_n$ and zeros at $Q_2, ..., Q_n$. Then the answer to the original problem is $f*f'$.

We give below a definition of this algorithm, and note that, for a constant curve and constant places over which the divisor lies, its running time is (approximately) proportional to (number of places) * $\log_2$ (largest multiplicity).

[#] I am very grateful to Professor Sir Peter Swinnerton-Dyer for suggesting this approach.

[*] If it exists - see Steps [2.2.1] and [2.4.1] below, and the discussion after the algorithm.

DIVISOR_TO_FUNCTION

Input: F(X,Y): The equation of an algebraic curve

(possibly in a multivariate representation).

D: a divisor of degree 0 on that algebraic curve.

Output: FUN: a function on that curve

which has that divisor as its divisor of zeros and poles, or FAILED if there is

no such function.

[1.1] POLES:= All the poles in D, with appropriate multiplicities.

ZEROS:= All the zeros in D, with appropriate multiplicities.

[1.2] FUN:= 1.

In FUN we accumulate the answer as we reduce the divisor D.

[2] While (number of ZEROS) > 1 do:

[2.1] P1:= Any member of POLES.

P2:= Any member of (POLES − P1).

Note that P1 and P2 could be the same, if a pole occurs in POLES with

multiplicity > 1. In theory it does not matter which elements of POLES we

choose, but in practice the running time of the algorithm is decreased if we

choose P1 and P2 such that the divisor P1+P2 divides POLES as often as

possible, and this can be accomplished by letting P1 be the element of

POLES of greatest multiplicity, and P2 that element of POLES' of greatest

multiplicity, where POLES' is POLES with the multiplicity of P1 halved

(rounding down).

[2.2] V:=COATES(F(X,Y),−P1−P2).

[2.2.1] If V = FAILED

Then go to [4].

[2.3] Z:=an element of ZEROS

> Here again it does not matter which element we choose, but the running
> time of the program is improved by choosing an element of greatest mul-
> tiplicity.

[2.4] V:= That linear combination of elements of V with a zero at Z.

[2.4.1] If V = FAILED

> Then go to [4].

[2.5] Z':= The other zero of V.

[2.6] While (P1+P2) | POLES and Z | ZEROS do:

[2.6.1] POLES:=POLES−P1−P2.

[2.6.2] ZEROS:=ZEROS−Z.

[2.6.3] If Z' | ZEROS

> Then ZEROS:=ZEROS−Z'
> Else POLES:=POLES+Z'.

[2.6.4] FUN:=FUN*V.

[3] If ZEROS is null

> Then return FUN.

[4] (Final step)

> We come here when we have some POLES left that we cannot explain away
> in pairs by the operation of [2.6]. This could occur either because there is
> only one element of POLES left (the usual case), or because one of [2.2] and
> [2.4] returned a FAILED (the exceptional case). The exceptional case is
> discussed at greater length below.

[4.1] V:=COATES(F(X,Y),−POLES).

4.2] If ZEROS = zeros of V

    Then return FUN*V

    Else return FAILED.

As mentioned at [4], there is no need for the intermediate elements of FUN to exist at all, even when the final problem* has a solution. This problem occurs, basically, because there might be no functions with only 2 poles (though this can't happen on curves of genus 1). We can continue by using the ordinary COATES procedure, as outlined in steps [4.1] and [4.2], but perhaps there is scope for a more enlightened process of adding poles, one at a time, until the space of functions is 2-dimensional, then taking a combination of these functions to kill a zero from the divisor, and repeating, but I have no idea how well this would behave in practice. It is quite clear that this chapter on Coates' algorithm has barely scratched the surface of what should become a fruitful area for further investigation by computer algebra workers.

Coates' Algorithm & Differentials

We can apply Coates' Algorithm to differentials, as well as to functions. We can write a differential as $F(X,Y)dX$ for some function F, and we need then to consider what our local parameters are at each place on the curve. There are three possibilities for the point over which our place lies:

1) The place could lie over infinity.

---

* The following example (for which I am grateful to Professor M.F. Singer) shows this. Let C be a curve of genus $g \geq 3$ which is not hyper-elliptic (i.e. cannot be represented as a 2-sheeted cover of the projective line, which essentially means that it is not of the form $y^2 = p(x)$). Let f be a rational (non-constant) function on C with the minimal number k of zeros and poles. Then, since C is not hyper-elliptic, it follows that $k>2$ (see any standard book on algebraic curves). Hence there is no rational function on C with only two poles, and an application of DIVISOR_TO_FUNCTION to (f) would fail.

In that case our local parameter is $t = (1/X)^{1/k}$ for some integer k, so $dt/dX = t^{-(k+1)}$, so our differential will have order $k+1$ less at infinity than the function F does.

2) The place could lie over a point a which is a root of Y.

In this case our local variable is $t = (X-a)^{1/k}$ for some integer k. This gives us $dX/dt = t^{k-1}$, so the differential has order $k-1$ more than the function.

3) The point can have neither property.

In this case $(X-a)$ is a local parameter, and the differential has the same order as the function.

Thus, in order to find a basis for the K-space of differentials with poles no worse than a specified divisor, we only need to adjust the multiplicities in the divisor at the (finite number of) places in categories 1 and 2 above, and then apply Coates' Algorithm to find a basis for the functions $F(X,Y)$.

In particular, we can find the genus of an algebraic curve, since it is the dimension of the space of differentials of the first kind, i.e. those with no poles. The divisor obtained from the zero divisor by applying the manipulations described in 1) and 2) above is termed the *canonical divisor* of the curve.

Implementation

Since Coates' Algorithm is going to be a fundamental building block in our integration algorithms, it is going to be worth our while to consider the implementation of this rather carefully. One particular technique of importance is the minimisation of the number of Puiseux expansions calculated in the course of Coates' Algorithm. In particular we need not recompute the Puiseux expansion of $A+B$ from scratch if we know the Puiseux expansions of A and B, but we can use the identity which says that the taking of Puiseux expansions commutes with addition.

During the operation of the algorithm INTEGRAL_BASIS_REDUCTION, it may happen that one entire row of the matrix A(I,J) is zero. If the I-th row is zero, then we can immediately deduce that V(I) should be replaced by V(I)/(X−P), and we can do this and recompute the I-th row of the matrix. Furthermore we need not re-expand this new function about the different places, but we can take the original Puiseux expansions and simply shift the power series by one term in order to produce the Puiseux expansions of the new V(I). This is especially suitable given our method for computing Puiseux expansions described in Chapter 2, since new terms can be computed at any stage while still making use of all previously computed partial results.

We may, having tried Coates' Algorithm on the divisor kD, wish to try it on $(k+1)D$. A naive application of the Algorithm COATES above (not using DIVISOR__TO__FUNCTION) would start each time from the same initial basis in step [1] of Algorithm COATES, thus throwing away in the computation of $(k+1)D$, all the information gained from the computation of kD. In fact we can start step [2] with the normal integral basis from step [4] of the kD attempt, since this basis is a set of n linearly independent members of M'. The advantage of this is that these members are much closer to having poles specified in $(k+1)D$ than the elements that would be computed in step [1], in particular they have the poles of kD. If D has total degree d, then the number of poles (counted with degree) that have to be introduced into the basis in the computation of Coates' Algorithm for D, $2D$,...,kD is kd rather than the $k(k+1)d/2$ that the obvious algorithm would require. There is an illustration of this in Chapter 5, under the heading Cayley's method, where it is shown that (using COATES rather than DIVISOR__TO__FUNCTION) a saving of 5.3 seconds can be made on a time of 38.2 seconds by this technique.

If we are using DIVISOR__TO__FUNCTION, then we can restart at the end of step [3] and add the poles and zeros of D, which is the difference between kD (whose function we had almost finished computing) and $(k+1)D$ (whose function we wish to compute).

Conclusions

There are two main application areas for the algebraic geometric treatment of expressions outlined in this chapter and the previous one. The first is the integration of algebraic functions, which is the main subject of this work. The second is concerned with locating integer (or rational number) solutions to sets of algebraic equations. Much work has been done on this (see the survey Swinnerton-Dyer,1976), but so far it has been approached by ad hoc arithmetical techniques, with all the algebraic geometry translated into FORTRAN (say) by hand.

We have raised the problem of finding all the functions with a specified set of poles, which is trivial for rational functions, and described Coates' Algorithm, which can provide a solution to it. This then leads to answers to the questions "does this divisor correspond to a function", "what is the genus of this curve" and "what are the differentials of the first kind". Later we will see that these are precisely the questions that are asked in the process of integrating algebraic functions.

Risch's Theorem

Introduction

Integration is probably the area of computer algebra with the greatest disparity between users' expectations and current systems' performance. Although heuristic techniques for integration are well developed (Moses,1967), there are still many advantages in having a decision procedure which can, for example, prove that an expression is unintegrable (Moses,1971).

This chapter describes an underlying body of theory to the area of finding (or proving non-existent) the elementary integrals of algebraic functions, where a function is *algebraic* if it can be generated from the variable of integration and constants by the arithmetic operations and the taking of roots of equations*, possibly with nesting (see Chapter 2 for a detailed discussion of such expressions and how we choose to represent them). By *elementary* we mean generated from the variable of integration and constants by the arithmetic operations and the taking of roots, exponentials and logarithms, possibly with nesting (this notion is made precise below). As a particular case of this problem, we have the question of deciding when an integral which at first sight appears to be elliptic can in fact be expressed in terms of elementary expressions. This problem was long thought to be insoluble, and Hardy summarised the classical position when he said

"No method has been devised whereby we can always determine in a finite number of steps whether a given elliptic integral is pseudo-elliptic, and integrate it if it is, and there is reason to suppose that no such method can be given". (Hardy,1916 pp. 47-8).

---

* The theory does not require that these roots should be expressible in terms of radicals. As mentioned in chapter 2, the implementation is currently restricted to algebraic quantities expressible in terms of square roots.

We shall explain a new algorithm for this problem, based on techniques from algebraic geometry (see the previous chapters), and discuss the implementation of this and the many open problems that remain in this area. We shall use the terminology and ideas of the previous chapters without comment. Note that this approach is different from Ng(1974), where the concern was with finding canonical forms for non-elementary elliptic and hyper-elliptic integrals. It is not certain quite what the relationships between these two techniques for attacking different but related problems are.

The problem "can we integrate this function?" might appear simple to pose, but this is not the case. In particular, suppose that we have an algorithm for deciding if algebraic functions over K are insoluble, where K is the field generated from the integers, log 2 and π by means of the operations of addition, subtraction, multiplication and division, and the functions $x \Rightarrow \log x$, $x \Rightarrow \exp x$, $x \Rightarrow |x|$. Then let $f(x)$ be any unintegrable function (e.g. $\sqrt{(1-x^2)(1-kx^2)}$ with k neither 0 nor 1) and A any expression which cannot be determined to be identically 0 or not (such expressions exist by the work of Richardson (1968)). Then consider whether $Af(x)$ is integrable: clearly an insoluble question. This is similar to Risch's exposition (Risch, 1969, Proposition 2.2), and the reader is referred to his work for further explanations of computability. Risch's work shows that we need a firm theoretical base for our discussion of integration, and this is provided by the subject of Differential Algebra, which we now describe.

Differential Algebra

Before we can discuss the problems of integration, we require a definition of what we mean by integration and integral. The obvious answer, and the definition we adopt, says that $F(x)$ is the integral of $G(x)dx$ iff $G(x)$ is the derivative of $F(x)$ with respect to x. This merely converts the problem from the definition of integration into the problem of differentiation. The study of differentiation as an algebraic operation (rather than an analytic one) is the province of *Differential Algebra,* a field of Mathematics founded by J.F. Ritt, although elements of it go back to Liouville and Laplace. We give here a very brief summary of elementary differential algebra in order to introduce the terminology we will need for our studies of differentiation and integration. For further details see the introduction to the subject by Kaplansky (1957), the survey by Ritt (1950) or the unified exposition of differential algebra by Kolchin (1973).

By a *differential field* we shall mean a field* together with a family D_i of unary operators (written as $a \rightarrow Da$) on the field (the *differentiations* of the field) satisfying:

$$D(a + b) = Da + Db,$$

$$D(ab) = aDb + bDa$$

for all a and b in the field and for each differentiation D. We define K to be a *differential extension* of L iff K is a field extension of L and every differentiation of K can be regarded as a restriction of a differentiation of L. Where there is only one differentiation D, we sometimes write a' for Da.

Any element which has image 0 under all differentiations is said to be a *constant* of the field. We obtain immediately that $D0=0$, and then that $D1=0$, so that any integer is a constant in the sense just defined. Except in chapter 6 we will only be dealing with one differentiation, namely the one that is the opposite to the integration we wish to perform. If X is the variable we wish to integrate with respect to (called *the variable of integration*), then we can convert $\{K(X,Y) \mid F(X,Y) = 0\}$ into a differential field by defining a' to be 0 for all a in K, $X'=1$ and $Y' = -(F_X/F_Y)$, where by F_X we mean the partial derivative of F with respect to X done purely formally (i.e. the derivative of cX^n is cnX^{n-1} and this partial differentiation is an additive operation). More generally, e.g. in the case of multivariate representations, if $F(X,Y,Z,..,W) = 0$, then $0 = F' = X'F_X + Y'F_Y + Z'F_Z + ... + W'F_W$, and this determines W' in terms of $X,Y,Z,...,W$. This clearly makes the field into a differential field, and integration is defined to be the inverse operation to differentiation where it is defined.

---

* Of characteristic 0 unless otherwise stated. "Ritt himself had no use for fields of non-zero characteristic and referred to them as 'monkey fields' " (Kolchin, 1973, p. xiii). I hope to be able to show later (see Chapter 8 in particular) that these fields are of some use in the theory of integration.

If K is a differential field and x,y belong to K with y non-zero, and $Dx=(Dy)/y$ for each differentiation D of K, we say that x is a *logarithm* of y, or that y is an *exponential* of x. Note that this agrees with the standard (analytic) definitions, in that the equation $Dx=(Dy/y)$ is normally taken as the definition of the exponential function of classical calculus, and that logarithms are defined as the inverse of exponentials. A differential extension field L of K is termed an *elementary extension* of K iff it is of the form $K(t_1,...,t_n)$, where for $1 \leq i \leq n$ we have one of:

t_i is a logarithm of an element in $K(t_1,...,t_{i-1})$

t_i is an exponential of an element in $K(t_1,...,t_{i-1})$

t_i is algebraic over $K(t_1,...,t_{i-1})$.

This definition makes precise the concept of an elementary function, in that it is one that can be expressed as a member of an elementary extension of a constant field.

Risch's Theorem

Just as the integral of a rational function is a rational function plus the sum of logarithms with constant coefficients, so the integral of an algebraic function, if it is elementary, is the sum of an algebraic function and logarithms of algebraic functions, these logarithms having constant coefficients. We shall say that such an integral consists of an *algebraic part* and a *logarithmic part*. An integral with no logarithmic part is said to be *purely algebraic* and one with no algebraic part is said to be *purely logarithmic*.

The first important result about the form of these integrals is Laplace's Principle (Hardy, 1916 pp. 9-10), which says that the integral of an algebraic function can be expressed so as to contain only those algebraic quantities which occurred in the integrand. This result can be proved by "elementary" means and Hardy gives a sketch of how to do so (pp. 36-42 and 46). However this is an easy consequence of our general work, see Corollary 2 below.

When one integrates a rational function, the logarithmic part comes precisely from the integration of terms such as $1/(X-k)$ in a partial fraction decomposition of the integrand, and the algebraic part comes precisely from the integration of all the other terms. This remark was generalised in a precise manner to algebraic functions by Risch, who stated the following Theorem (1970):

"Let w be a differential in $\{K(X,Y) \mid F(X,Y) = 0\}$. Let $r_1,...,r_k$ be a basis for the Z-module generated by the residues of w. Thus, at each K-place P, w has residue $\sum_{i=1}^{k} a_{iP} r_i$ with a_{iP} in Z. Let the divisor d_i be given by multiplicities a_{iP} at the places P. Then, iff $\int w$ is elementary, there are $v_0,...,v_k$ in $K(X,Y)$ and integers $j_1,...,j_k$ such that d_i to the power* j_i is the divisor of the function v_i and

$$w = dv_0 + \sum_{i=1}^{i=k} \frac{r_i}{j_i} \frac{dv_i}{v_i}$$

i.e.

$$\int w = v_0 + \sum_{i=1}^{i=k} \frac{r_i}{j_i} \log v_i".$$

In this theorem, v_0 is the algebraic part of the answer, and the remainder is the logarithmic part, which depends only on the residues of the integrand. This is a straightforward generalisation of standard integration methods for rational functions, except that there is the possibility that the divisors may be rationally equivalent to 0, instead of linearly equivalent. If they are, then j_i is that integer multiplier which makes d_i into the divisor of a function. This theorem immediately gives rise to a possible algorithm for integrating algebraic functions:

---

* in the sense of multiplication of divisors introduced in Chapter 3; i.e. with all the multiplicities multiplied by j_i and the places unchanged.

RISCH__ALGEBRAIC

Input:

F(X,Y): the equation of an algebraic curve

f(X,Y): a function of X and Y

(w will then be $f(X,Y)dX$)

Output:

I : The integral of f(X,Y)dX,

(or NOT__ELEMENTARY if there is no elementary integral)

[1] POTENTIAL__POLES:= "INFINITY"

combined with all factors of the denominator of f(X,Y)

(which is a polynomial in X by the remarks on representations of algebraics contained in Chapter 2).

[2] RESIDUES:=for each U in POTENTIAL__POLES

for each place V lying over U

Collect (V,Residue of f(X,Y) at V)

[3] Z__BASIS:=basis for the Z-module of the residues formed in step [2]

[4] For each Ri in Z__BASIS do

[4.1] Di:= for each U in RESIDUES collect:

(Place.coefficient of Ri in residue).

This corresponds to the divisor d_i in Risch's Theorem.

[4.2] Ji:=FIND__ORDER Di.

By Risch's theorem, the integral can only exist if the divisor Di is of finite order. This tells us the order of the divisor Di, i.e. which is the least power of it which is linearly equivalent to 0. If there is none such, then this returns INFINITE.

[4.3] If Ji = INFINITE, then return NOT__ELEMENTARY.

[4.4] INTEGRAL:=INTEGRAL+
    (Ri/Ji)*log(DIVISOR__TO__FUNCTION(F(X,Y),Di**Ji))

> We have now found all the logarithmic parts of the integral, which we can now remove from the integrand in order to find the algebraic part.

[5] Now find the algebraic part of the answer.

[5.1] f(X,Y):=f(X,Y)−derivative of INTEGRAL.

[5.2] ALG__PART:=FIND__ALGEBRAIC__PART f(X,Y)

[5.3] If ALG__PART = NOT__ALGEBRAIC
    then return NOT__ELEMENTARY.

> In this case, we could find all the logarithmic parts of the integral, but not the algebraic part.

[6] return INTEGRAL + ALG__PART.

This is, in fact, a perfectly workable algorithm, subject to the definition of the subsidiary algorithms. Many of these, although their precise description may appear strange, are perfectly familiar algorithms of computer algebra. The factorisation in step [1] is merely in $K[X]$ because of the remark in Chapter 2 which says that we can represent an element of $\{K(X,Y) \mid F(X,Y) = 0\}$ as a polynomial in X and Y divided by a polynomial in X. The finding of the basis for the Z-module in step [3] can be re-expressed as asking which rows of a matrix are linearly independent, and this is easy to answer, although not easy to answer efficiently for large matrices. If this matrix is large, though, the integrand which gave rise to it is almost certainly far too large to integrate anyway, so this point should not concern us much.

The problem of finding the residue at a place requires the computation of the t^{-1} term in the Puiseux expansion (where t is a local parameter at the place in question). As described in Chapter 2, Puiseux expansions are really just Laurent expansions, and can be computed by the techniques described in Chapter 2, based on those in Norman (1975). Although a significant body of code is required to perform these expansions correctly and efficiently, the difficulties involved are technical rather than mathematical.

The algorithm DIVISOR__TO__FUNCTION is described in Chapter 3. FIND__ORDER is harder, and we will return to it in a later Chapter.

Proof of Theorem

Risch enunciated the theorem mentioned above as a consequence of his major theorem on integration (the proof of which was not included in the published work, though a proof can be deduced from (Risch, 1969, pp. 171 et seq)). It is therefore necessary to prove this theorem before we rely on it for the construction of a theory of integration. The theorem is stated as an "if and only if" theorem, but if w has the form stated, then it is clearly elementary. Hence the major task is to prove that all elementary differentials are of this form. Rather than adapt Risch's earlier proof (which was mostly concerned with transcendental functions), we shall give a new short one, relying on a recent result of Rosenlicht (1976).

We can observe intuitively that differentiation does not remove exponentials, or logarithms unless they occur as ... + $c \log f(x)$. Furthermore differentiation does not remove algebraics. Therefore it seems intuitively reasonable that the integral of an algebraic function, if it is elementary, can be expressed in the form $u(x) + \Sigma c_i \log u_i(x)$, where the c_i are constants. This remark was made by Laplace, and first proved, by an ingenious descent argument, by Liouville(1833c). This proof can be found in Ritt's book on Liouville's work (Ritt, 1948, p.20). We will base our attack on the more advanced methods of abstract algebra, since it seems likely that any generalisation will have to come from this source.

Lemma 1 For $f(X)$ in $L = \{K(X,Y,Z,...) \mid F(X,Y) = 0 \quad G(X,Y,Z) = 0...\}$ with K algebraically closed, $\int f dX$ is elementary in the sense that there is an elementary extension of

L containing an element g with $f=g'$, iff there are constants $c_1,...,c_n$ linearly independent over the integers in L and elements $u_1,...,u_n,v$ in L such that $f = v' + \Sigma c_i\dfrac{u'_i}{u_i}$.

Proof: Apart from the part about linear independence of the c_i, this is a restriction of Theorem 3 of Rosenlicht(1976) to the case of one differentiation of a special form. And if the c_i are not linearly independent over the integers, we can combine terms to make them so, using the identity $D(u^k)/(u^k) = kDu/u$.

Corollary 2 The elementary integral of an algebraic function can contain no algebraics depending on the variable of integration other than those which occur in the integrand[*].

Proof: If the integral is elementary, then it has the form specified in the Lemma above, and so its integral is $v + \Sigma c_i \log u_i$ where v and the u_i belong to the differential field in which the original integrand lay, and hence are expressible in terms of the same algebraics.

Note the importance of the phrase "depending on the variable of integration" in the above statement of Laplace's Principle. The Principle is not true without this restriction, as can be seen from Risch (1969, Proposition 1.1) who considers the integral of $1/(X^2-2)$, which necessarily contains $\sqrt{2}$.

Lemma 3 The field of power series in one variable $K((t))$ is a differential field with $\Sigma a_i t^{i'} = \Sigma iat^{i-1}$.

Proof: Addition is obvious, and multiplication follows from comparing the left and right hand sides.

---

[*] This is Laplace's Principle (Hardy, 1916, pp. 9-10) quoting Laplace, "L'intégrale d'une fonction différentielle (algébrique) ne peut contenir d'autres quantités radicaux (sic) que celles qui entrent dans cette fonction" [I have been unable to find precisely this form of words in any of Laplace's works, but I did discover the following: "l'intégrale d'une fonction différentielle ne peut renfermer d'autres quantités exponentielles et radicales que celles qui sont contenues dans cette fonction" (Laplace, 1820, livre I, p. 5).] Note that Laplace did not state explicitly that this only relates to algebraics depending on the variable of integration. Another proof can be found in Ritt's work (1948, pp. 28-31).

Lemma 4 If x is the variable of differentiation, and $dz=y\,dt$, then $y=z'$, where $'$ is the "power-series" differentiation of the previous lemma.

Proof: This is Theorem 8 of (Lang, 1957, p.247).

Corollary 5 If t is a local variable, the Puiseux expansion of $f'\,dx/dt$ is the derivative (in the sense of Lemma 3) of the Puiseux expansion of f.

Corollary 6 If F has a purely algebraic integral, and the Puiseux expansion of F is $\Sigma\,a_i t^i$, then the Puiseux expansion of its integral is $\Sigma\,\dfrac{a_i t^{i+1}}{i+1}$.

Lemma 7 The Puiseux expansion of the derivative of $\log F(X)$ is that of $F'(X)/F(X)$, in particular, if $F(X)$ has a zero (or pole) of order n at the place P, then $(\log F(X))'$ has a residue of n at this place.

Proof: The first part follows from the definition of logarithms given earlier, which characterises them by precisely this property. The second part follows from a consideration of Puiseux expansions: F has Puiseux expansion $a_n t^n + .$ and F' has expansion $n a_n t^{n-1} + ...$, giving a quotient of $nt^{-1} + ...$, as required.

Proof of Theorem. If w has an expression of the form stated, then it is clearly integrable. Conversely, suppose w is integrable. Then by Lemma 1, we can write w as $v dx + \Sigma\,c_i(\log u_i)'dx$, and we can assume that none of the u_i are constant, for if they were then they cannot contribute to the derivative of the integral. Choose such an expression with n minimal. Now, by Lemma 7, the residue of w at P is $\Sigma\,c_i b_{iP}$, where b_{iP} is the order of u_i at P. Also, since each u_i has at least one pole and the c_i are linearly independent, each c_i appears in at least one residue of w. Hence the c_i form a basis for the set of residues of w (or possibly of a larger module, but certainly one which only includes fractional multiples of the residues). So, if r_i is our basis for the set of residues of w, considered as a Z-module, then there are integers j_i such that the set $\{r_i/j_i\}$ is a basis for the set of c_i, so that we can write c_i as $\Sigma\,\dfrac{n_{ij}r_j}{j_j}$. And then we can re-express the sum of logarithms to have coefficients r_i/j_i and the theorem is proved.

Corollary 8 An integrand with no residues can have no logarithmic part in its integral.

The Algebraic Part

At step 5.2 in the integration algorithm outlined above, we are left with a function to integrate whose integral, if it is elementary, is purely algebraic. This is a much easier problem than the general one, and was considered, and essentially solved, more than 150 years ago (Liouville, 1833a, 1833b). We describe here a way of solving this problem which interacts closely with the rest of this work, and which is frequently very efficient since many of the computations have been made in the process of finding the logarithmic part of the integral. There are several ways in which this problem can be solved, and the one described here is not necessarily the most efficient,[#] it is merely the one which is closest to our algebraic-geometric view of the problem of integration.

This algorithm is based on the observation that the poles of an algebraic function $f(X,Y)$ and its derivative $f'(X,Y)$ are closely related (see Corollary 6 above). In fact, where f' has a pole of order n, f has a pole of order $n-1$ (or $n+1$ if the place lies over infinity). This is a generalisation of the usual rule for integrating powers of X, and can easily be proved by appealing to Puiseux expansions. We cannot say much about the zeros of f, because the Puiseux expansion of f could have a constant term which will disappear on differentiating.

There is one difficult point in this process. We cannot completely determine the integral this way, since the integral is indeterminate up to a constant of integration. Initially I tried various strategies for determining whether or not the indeterminacy in the linear combination of the functions with appropriate poles was due to this constant of integration or not, but I could not find a reasonable method of doing this. I therefore adopted the strategy of choosing a point P (not a pole of the integrand) and demanding

---

[#] In the case of simple radical extensions, the theory simplifies substantially (and is essentially due to Chebyshev (1853)). Trager (1979) has recently published an algorithm for finding the algebraic part in this case which appears to be very efficient, since it only relies on polynomial algorithms, but I have been unable to find an implementation of this in order to obtain some comparative timings.

that the algebraic part of the integral have the value 0 there, i.e. making a definite choice
for the value of the constant of integration. This leads to some slightly curious integrals at
times, but there is clearly scope for a heuristic post-pass to choose more "appealing" values
of the constant of integration after the complete integral has been found.

FIND__ALGEBRAIC__PART

This algorithm is stated for the simple cases; there are complications when $(X-a)$ is not a
local parameter near $X=a$, but the principle is the same.

Input:

　　F(X,Y): the equation of an algebraic curve

　　f(X,Y): a function of X and Y

　　　　　(w will then be $f(X,Y)dX$)

Output:

　　FUNCTIONS : The integral of $f(X,Y)dX$,

　　　　　(or NOT__ALGEBRAIC if there is no algebraic integral)

[1] POTENTIAL__POLES:= "INFINITY"

　　and all factors of the denominator of f(X,Y) (which is a polynomial in X).

[2]

　　　DIVISOR:=for each U in POTENTIAL__POLES

　　　　　　　for each place V lying over U

　　　　　　　Collect (V,order of f(X,Y) at V).

　　　This order is computed from the Puiseux expansion of f(X,Y) at V.

[3] For each U in DIVISOR

　　　　ORDER:=minimum(0,ORDER+1).

Or ORDER-1 if the place lies over infinity. The $+$ and $-$ signs have changed since the description of the theory above, since the order of a pole is always negative.

[4] FUNCTIONS:=COATES(F(X,Y),DIVISOR).

> We actually need a version of COATES which will deal with places at infinity, but, as explained in Chapter 3, this is not particularly difficult.

[5] For P = 1,2, ... do:

> if there is no element of DIVISOR lying over P
>
> > then go to 6.

> > This loop must eventually terminate, because DIVISOR is a finite list of poles. P is then the point at which we can insist that the integral must have the value 0, in order to fix the value of the constant of integration.

[6] Let A[i] be the value of FUNCTIONS[i] at P.

> Eliminate one of the FUNCTIONS
>
> > by using the constraint $\Sigma A[i] = 0$.

[7] For each U in DIVISOR with ORDER$\neg$=0:

> For each At^{-k} term in the Puiseux expansion of $f(X,Y)$:
>
> Ensure that the integral has an $-At^{-k'}/k'$ term, where k' is $k+1$ (or $k-1$ if J lies over infinity). This is done by applying a linear constraint to FUNCTIONS as produced in step 4, and in practice these linear constraints are not dense for positive k. We cannot do this for $k'=0$.

[8] When FUNCTIONS is reduced to one possibility,

> differentiate it, and then we have $f(X,Y)$ (in which case this is the integral) or not, in which case $f(X,Y)$ has no algebraic integral, and the answer is NOT__ALGEBRAIC.

This algorithm is sufficient to prove that many standard integrands are unintegrable, e.g. $1/\sqrt{(x^2-1)(x^2-k)}$, which has no poles.

A Partial Algorithm

Even without an effective realisation of FIND__ORDER the above techniques give us a partial algorithm for integrating algebraic functions: partial in the sense that if the algorithm is presented with an integrable algebraic function, then it will terminate with the integral; if the function is not integrable then it may return NOT__ELEMENTARY or it may run for ever.

We recall that the order of a divisor on an algebraic curve is defined as the smallest power of it such that there is a function corresponding to it, i.e. such that Coates' Algorithm succeeds on it. Hence we can produce a place holder for the FIND__ORDER procedure which works by trying each power of Di in turn until it finds such a power. This procedure will of course run for ever if the divisor is rationally inequivalent to zero.

This approach does demonstrate that we need only have a bound on the order of a divisor, rather than its actual order. If we assume that all divisors have order 1, then we will be correct for curves of genus 0, i.e. those integrands that could be solved by a trigonometric substitution. The more elaborate order-finding processes discussed below are a reflection of the difficulty of identifying elementary integrals among apparently elliptic and hyper-elliptic integrals.

We note that this partial algorithm is complete for all integrals that do not contain a logarithmic part, since these integrands will have no residues.

Efficiency Points

This algorithm is not necessarily efficient, even in the case of integrals which correspond to divisors of order 1. For example, consider integrating the function

$$\frac{1}{\sqrt{x^2+1}} + \frac{100}{\sqrt{x^2+10000}}.$$

This has residues of $99, 101, -99, -101$ at the four places lying over infinity. The corresponding function found from the call of DIVISOR_TO_FUNCTION in step 4.4 of the algorithm is

$$\frac{x^{101}}{(\sqrt{x^2+1}-1)(\sqrt{x^2+10000}-100)^{100}},$$

and, as explained in the previous chapter, finding this function by naive methods would involve the solution of a 200 by 200 set of dense linear equations. Even using the method of DIVISOR_TO_FUNCTION, we still need to compute the answer from DIVISOR_TO_FUNCTION, and its denominator is a dense even polynomial of degree 100 in X. Fortunately this function does have structure and can be computed in a non-expanded form, so this is not as much of a problem as might appear. Some examples on these lines are quoted as Example 8 in Appendix 2. Integrating each term separately involves almost no work, however, so it is clear that we should do this wherever possible, for we shall reduce the average running time of the algorithm, even when we do not decrease the maximum running time. In other words, there is significant scope for heuristics to improve the performance of the system, even if they do not increase the range of soluble problems.

An earlier version of the program described in this monograph, and the transcendental integration package of Norman & Moore (1977), have been incorporated with a pattern matcher and set of rules for user defined integrals into the algebra system REDUCE by Harrington(1977, 1979b), and his techniques, which rely on ours or Norman & Moore's as appropriate, are probably the best approach to a practical integration system, since they combine the elegance and certainty of an algorithmic approach with the speed of a semi-heuristic approach.

The Problem of Torsion Divisors

Introduction

This chapter and the next three are concerned with the theory and practice of the FIND-ORDER procedure, which, as we saw in the last chapter, is a necessary part of our integration algorithm, and which turns out to be the mathematically the most difficult. This chapter will outline the general nature of the problem, with special reference to the simplest non-trivial case, viz. problems involving the square root of one cubic or quartic[#] and involving no constants other than the rationals.

The simplest case of all is clearly that with just one square root of a linear of quadratic expression involved. In this case[$] we have that the procedure FIND-ORDER can always return 1, because all divisors of degree 0 are linearly equivalent to 0. This remark is equivalent to saying that such a curve can always be rationalised, and hence the problem is reducible to that of integrating a rational function. This remark implies that our algorithm, with FIND-ORDER written so as to yield at least 1 (though not necessarily the correct answer), will solve all the "obvious" cases that any heuristic algorithm would solve.

Unfortunately, producing a complete FIND__ORDER procedure is substantially more difficult than anything we have done so far, and involves much more mathematics. One simple example of this is given in (Appendix 2, Example 3), where we have a divisor of order 3 from an integration involving $SQRT(X^3+1)$, which appears little more complicated

---

[#] these are (in general) curves of genus 1, or elliptic curves, and the integrals defined over them are the elliptic integrals, which provide the simplest algebraic example of non-integrability.

[$] known as a curve of genus 0 (curves of genus 0 may also have more complicated expressions, but they can always be reduced to this form).

than $SQRT(X^2+1)$. We will need to invoke a substantial body of algebraic geometry and number theory, and several recent results, in order to understand this problem better.

Informally, our problem is the existence of divisors which are not divisors of functions, but which have a multiple which is the divisor of a function. Given Coates' Algorithm (Chapter 3), we can decide whether a divisor is the divisor of a function or not, so let us identify two divisors if (and only if) their difference is the divisor of a function. Then our problem is reduced to deciding whether a divisor (of degree 0) does or does not have a multiple which is (identified with) 0. A divisor which does is termed a *torsion divisor*.

We must consider all kinds of divisor on algebraic curves, because they are all relevant to the problem of integration[#].

To express these necessary mathematical foundations in more formal terms, we proceed as follows: The divisors which are linearly equivalent to 0 form a subgroup of the group of all divisors, and we have, in Coates' algorithm, a procedure for deciding if a divisor lies in this subgroup. Consider now the group of all divisors of degree 0, quotiented out by this subgroup. This group, which we shall call the *Jacobian divisor group* of the curve, is an Abelian group, which is finitely generated[*]. It can therefore be regarded as the direct product of a number of infinite cyclic groups and a finite Abelian group (known as the *torsion part* of the group). A divisor is rationally equivalent to 0 iff its image lies inside the torsion part of the curve, and then its order as a divisor is equal to its group-theoretic order as a member of this Abelian group. The size of the finite portion of the Jacobian divisor group is known as the *torsion* of the group, or of the curve, and it therefore follows that the order of any divisor which is rationally equivalent to 0 must divide the torsion of the curve.

[#] To every divisor there is a differential of the third kind having that as its residues (Lang, 1972, p. 44).

[*] by the Mordell-Weil Theorem (Mordell,1922), (Weil,1928) and (Lang & Neron, 1959), as applied to the Jacobian of the curve (see below).

Elliptic Curves

In this section we apply the above remarks to the special case of elliptic curves, i.e. curves of genus 1. We do this both because of the intrinsic importance of this case, subsuming as it does all the elliptic integrals, and because it serves as a simple illustration of the problems we will face later with curves of arbitrary genus. Much of this material is described by Cassels(1966) as updated by Tate(1974). We do this for two reasons: many practical problems lie on elliptic curves, including all those involving two square roots of quartics, or one square root of a cubic or a quartic; and there is a lot of special technology that can be applied to this case, but not the cases of higher genus (largely because elliptic curves have been much more extensively studied by mathematicians than curves of higher genus).

This study requires a number of results from algebraic geometry and from the study of elliptic curves which are presented here. The first set show how the group under discussion can be given a concrete geometric model.

Theorem 1 (Riemann-Roch)[#] If D is any divisor on a curve of genus 1, and $l(D)$ is the dimension of the K-space $L(D)$ of functions f with $D+(f)$ effective, then $l(D)$ is $0,1,$degree D depending on whether the degree of D is $<0, =0, >0$.

Lemma 2[*] On an elliptic curve with a point O distinguished, the linearly inequivalent divisors of degree 0 are in (1-1) correspondence with the points of the curve. Also O corresponds with the zero divisor.

Therefore we can create an addition on the points of the curve by making this correspondence into a group isomorphism. Then $A+B=C$ means that the divisor $A+B$ is linearly equivalent to the divisor $C+O$, or that $(A-O) + (B-O)$ is linearly equivalent to $(C-O)$.

See (Fulton, 1969, p.210) or (Chevalley, 1951) or any other work on Algebraic Geometry. Cassels(1966) takes this as his definition of a curve of genus 1.

* This is proved in Cassels, (1966, p. 211).

Theorem 3 [$] The elliptic curve E, with distinguished point O, is birationally equivalent[+] to the curve $Y^2 = X^3 - AX - B$ for some A and B, (the so-called *Weierstrass canonical form* for the curve) and O maps onto the point at infinity. Furthermore, on this elliptic curve, $P + Q + R = 0$ iff the three points P Q and R are collinear. Therefore (X_1, Y_1) and (X_2, Y_2) have sum

$$X = \frac{((-A + X_1 X_2)(X_1 + X_2) - 2B + 2Y_1 Y_2)}{(X_1 - X_2)^2} \quad \text{and}$$

$$Y = -\frac{(X(Y_1 - Y_2) + X_1 Y_2 - X_2 Y_1)}{(X_1 - X_2)}.$$

If $X_1 = X_2$ then these formulae collapse, and in fact if $Y_1 \neq Y_2$ then the point (X, Y) is at infinity, and if $Y_1 = Y_2$ then

$$X = \frac{(X_1^4 + 2AX_1^2 + 8BX_1 + A^2)}{4Y_1^2} \quad \text{and}$$

$$Y = \frac{Y_1 + (X - X_1)(3X_1^2 - A)}{2Y_1}.$$

We can regard problems of linearly inequivalent divisors of degree 0 as problems in this group of points. We can transform any divisor into a single point on a Weierstrass elliptic curve, and then ask questions about this point. Before we can apply the above theorem, we need to have an effective version of it, which is provided by the following algorithm (due to Baker & Coates (1970), see also Baker (1975)).

[$] Cassels (1966) Theorem 7.1 on p.211 and Lemma 7.1 and discussion on pp. 213-4.

[+] i.e. we can make rational transformations each way. As Cassels (1966) remarks, the traditional language is perverse inasmuch as a rational transformation need not have rational coefficients.

WEIERSTRASS_FORM

Input:

F(X,Y): An equation defining a curve of genus 1.

     There is in fact no need for this equation to be presented in this (primitive element) form; since the algorithm COATES will deal with a multivariate representation this algorithm will.

D: A divisor on the curve.

Output:

F'(X',Y'): An equation of the form $Y'^2 = X'^3 + aX' + b$

    which is birationally equivalent to the equation F.

D':   The divisor on F' which corresponds to the

    divisor D on F under a birational equivalence from F to F'.

[1] Let Q be a place at infinity on F(X,Y)=0.

[2] Let X_1 have a pole of order precisely 2 at Q, and no other poles, on F(X,Y)=0, by the use of Coates' Algorithm (Chapter 3).

[3] Let X_2 have a pole of order 3 similarly.

[4] Let (a,b,c,d,e,f,g) be a non-zero solution (in K) of the equation

$$a + bX_1 + cX_2 + dX_1{}^2 + eX_2{}^2 + fX_1{}^3 + gX_1X_2 = 0.$$

    Such a solution exists since we have 7 functions, each with divisor $\geq -6Q$, and so they cannot be linearly independent (by the Riemann-Roch Theorem).

    Now e cannot be 0, because

$$eX_2^2 + (cX_2 + gX_1X_2) + (a + bX_1 + dX_1^2 + fX_1^3) = 0,$$

and the first bracketed term either vanishes or has a pole of odd order at Q, and the second bracketed term either vanishes or has a pole of even order at Q. Then f is non-zero, because e and f correspond to the only terms with poles of order 6 at Q.

[5] $X' := X_1$

   $Y' := 2eX_2 + gX_1 + c$

   $A := -4ef$

   $B := g^2 - 4de$

   $C := 2cg - 4be$

   $D := c^2 - 4ae.$

This is an elliptic equation (since A is non-zero and it is birationally equivalent to the original curve of genus 1) which is satisfied by virtue of the equation in [4].

[6] $Y' := AY'$

   $X' := AX'$

   $C := AC$

   $D := A^2D.$

This is the effect of multiplying the whole equation by A^2 and then writing Y' for AY', X' for AX'. The coefficient of X'^3 is then forced to 1.

[7] $K := B/3$

   $X' := X' + K$

   $C := C - 3K^2$

   $D := D - K^3 - CK.$

This is the effect of writing $(X' + K)$ for X' in the equation above. We now have a genuine Weierstrass canonical form, and need only transform the divisor appropriately.

[8] $D' :=$ for each (N,P) in D collect:

$(N, X'(P))$

Mazur's Result

This mathematical technology immediately gives us an improvement on the partial algorithm described in the last chapter: It is now known (Mazur, 1978, Theorem 2; see also Mazur (1977) Theorem 1) that, in the restricted case of an elliptic curve (i.e. a curve of genus 1) defined[#] over the rationals, the torsion is bounded by 16. In fact a slightly stronger statement can be made: the maximum order of any element in the torsion part of the Jacobian divisor group of an elliptic curve defined[#] over the rationals is 12. This particular bound on the torsion gives a complete integration algorithm for curves of genus 1 over the rationals, and hence resolves this case of the problem posed by Hardy (Hardy, 1916 pp 47-8) and stated in the introduction in the case of elliptic integrals defined over the rationals.

It is a long-standing conjecture (attributed by Cassels to "the folklore" (Cassels, 1966, p. 264)) that, for all finitely generated[*] fields K, there is a bound, independent of the curve $\{K(X,Y) \mid F(X,Y) = 0\}$, on the torsion, but mathematicians have made little progress on this problem in general (since the consensus of opinion is that Demjanenko's (1971) results are not fully justified, and Cassels' review of them stated that "... Unfortunately, the exposition is so obscure that the reviewer has yet to meet someone who would vouch for the validity of the proof; on the other hand he has yet to be shown a mistake that unambiguously and irretrievably vitiates the argument").

[#] We mean by this that, not only does the equation of the curve have purely rational coefficients, but that the places which occur in the divisor in question have rational coefficients.

[*] There is an apparent contradiction here, since in Chapter 2 we said that we would consider only algebraically closed fields K. However, any particular problem contains only a finite number of algebraics, and so can be viewed in a finitely generated field. This remark requires proof but is essentially obvious, since if a divisor is of infinite order over a field, it is of infinite order over any algebraic extension of that field, and the whole problem (and answer) is a finite expression.

The paradigm proposed above, of bounding the torsion and then searching all multiples of the divisor up to this bound for one for which Coates' Algorithm can return a function, is unsatisfactory for practical reasons, though it does provide a complete solution to the integration problem. The following sections give some alternative methods which can supplement this general paradigm and greatly increase the efficiency of the system in relevant cases.

One very simple enhancement, which only applies to the case of elliptic curves over the rationals, is the Lutz-Nagell Theorem (see Chapter 7 for details, or (1978), p.55, Theorem 2.2). This states that a rational point on an elliptic curve in Weierstrass standard form, with integer coefficients is a torsion point only if it is actually integral[*] (i.e. X, and therefore Y, must be integers). This is clearly a very cheap test, once the curve has been reduced to canonical form.

Application of Mazur's work[#]

This section considers an application of the result of Mazur described in the previous section. The Mazur bound of 12 for the torsion of any individual point (or divisor) on an elliptic curve defined over the rationals can be used (in suitable cases) to determine whether or not a function is integrable. Here we note an alternative test for integrability, based on this bound, which does not rely on the construction of the integral if such exists. While this result is not directly related to the main thread of this work, it is a method of proving algebraic expressions unintegrable, and does rely on many of the techniques we have mentioned for its proof, though not for its implementation.

We consider the problem of integrating $1/f(x)$, where $f^2(x) = a_0^2 x^4 + a_1 x^3 + a_2 x^2 + a_3 x + a_4$, where the a_i are integers, although the integral of any

---

[*] I am grateful to Dr. N.M. Stephens for drawing my attention to this test in connection with the Mordell-Weil groups of elliptic curves.

[#] I am grateful to Professor A. Schinzel for drawing my attention to this work.

quartic with rational coefficients can be transformed into this form, and we require this polynomial to be square free.

Lemma. 4 If the continued fraction for $f(x)$ is periodic, then it becomes so immediately.

Proof: This follows from Schinzel (1962, para. 2).

Without loss of generality, we may assume that $f(x)^2$ is given by $x^4 + 6bx^2 + 4cx + d$. Then the following result follows from Schinzel(1962) quoting Halphen(1886, Vol I p. 120 and Vol II p. 591):

Lemma 5 The point $(-b,c)$ has order r on the elliptic curve $Y^2 = 4X^3 - (3b^2 + d)X - (bd - b^2 - c^2)$ iff the continued fraction for $f(x)$ has period $(r-1)$ or $2(r-1)$.

Theorem 6 The function $1/f(x)$ is integrable iff the continued fraction for $f(x)$ is periodic with period <23; $f(x)^2$ being a quartic with rational coefficients.

Proof: A point on an elliptic curve over the rationals has order at most 12, by Mazur's work, so in the Lemma above $(r-1) < 12$ and $2(r-1) < 23$. This proves the result, and provides us with a finite test for the integrability of $1/f(x)$.

More results about the inter-relation of continued fractions and integrability can be found in Chebyshev (1857 and 1860), and it appears that there is much more that could be done to relate continued fractions to algorithms for integration. Since the techniques of Trager (1979) allow us to compute the algebraic part of integrals in single radical extensions, and since Chebyshev's work (1857 and 1860) applies to this case also, there is clearly scope for using the two together. We should note that these continued fractions can be computed effectively by a suitable generalisation of the "Chain" method of Churchhouse (1976).

Cayley's method

In this section we consider the problem of discovering if a divisor on an elliptic curve is linearly equivalent to 0 without the necessity of using Coates' algorithm. This work is based on a recent paper (Griffiths & Harris, 1978) which restates several results of Cayley(1853 and 1861). We first note that our section above on Elliptic Curves implies that we need only consider single points on our elliptic curve.

Theorem 7 [*] Let p be a point on an elliptic curve E. We can rewrite E as $y^2 = (x-a)(x-b)(x-c)$ where x has the value 0 at p. Write y as $\sum_{i=0}^{\infty} A_k x^k$ (viz as a Puiseux expansion). Then p is of finite order n iff the relevant determinant below vanishes:

$$\begin{vmatrix} A_2 & A_3 & \cdots & A_{m+1} \\ A_3 & A_4 & \cdots & A_{m+2} \\ \cdots & \cdots & \cdots & \cdots \\ A_{m+1} & A_{m+2} & \cdots & A_{2m} \end{vmatrix} \qquad n = 2m + 1,$$

$$\begin{vmatrix} A_3 & A_4 & \cdots & A_{m+1} \\ A_4 & A_5 & \cdots & A_{m+2} \\ \cdots & \cdots & \cdots & \cdots \\ A_{m+1} & A_{m+2} & \cdots & A_{2m-1} \end{vmatrix} \qquad n = 2m.$$

Then we can apply this test before we use Coates' Algorithm on curves of genus 1, since this determinant evaluation is much cheaper than an application of Coates' Algorithm[#]. Although this does not necessarily decrease the computational order of our integration algorithm on curves of genus 1, it definitely supplies a large improvement in the actual running time for those cases which involve a consideration of the procedure

---

[*] Griffiths & Harris, (1978) equation (10), restating Cayley (1853, Theorem, p.376).

[#] In fact, since this determinant is of a special (Toeplitz) form, we can evaluate the determinant even more cheaply than might appear to be the case (Yun & Gustavson, 1979).

FIND__ORDER.

   Example. Let us consider the application of this improvement to the case of Tate's curve with $D=2$ (see Appendix 2 Example 4). This gives us the curve $y^2 = 4x^3 - 15x^2 + 8x + 16$, on which we wish to integrate the function $P(x,y)/Q(x,y)$, where

$$P(x,y) = (-14yx^5 - 30yx^4 + 251yx^3 + 60yx^2 - 688yx - 488y$$

$$+ 80x^6 - 320x^5 - 341x^4 + 1992x^3 + 448x^2 - 3200x - 1792)$$

and

$$Q(x,y) = 7x(4yx^5 + 5yx^4 - 51yx^3 - 4yx^2 + 112yx + 64y$$

$$-20x^6 + 79x^5 + 41x^4 - 368x^3 - 32x^2 + 512x + 256)$$

which gives rise to a divisor of order 7. In an attempt to integrate it without using Cayley's method, we have 7 applications of Coates' Algorithm, with individual CPU consumption of 1.9 seconds, 3.5 seconds, 6.5 seconds, 9.6 seconds, 16.5 seconds, 25.8 seconds and 32.9 seconds respectively. This gives a total time in Coates computations (excluding the computation of the genus of the curve) of 96.7 seconds in a total runnning time of 107 seconds. Re-running this example using Cayley's test for the order of points gives two Coates steps (because we test to see if the divisor is of order 1 before trying the FIND__ORDER procedure) of 1.9 and 38.2 seconds[$] in a total CPU consumption of 48.3 seconds.

[$] Note that, although we are considering the same divisor, viz one with multiplicities of 7 and −7, the time of 38.2 seconds in this case is different from the 32.9 seconds quoted for the case without Cayley's method. This is because in the case without Cayley's method, we have already computed a basis for the space of functions with poles & zeros of order 6, and we can use this as a starting point in Coates' Algorithm, while in the current case, we can only start from a basis with poles and zeros of order 1, and hence we have to perform more reduction steps. This matter is discussed under the heading "Implementation" in Chapter 3.

Jacobians

We can generalise some of the above concepts by introducing the Jacobian of an algebraic curve. We first need the concept of a (projective) *algebraic variety* , which we define as (Shafarevich, 1972, p.36) a subset of projective[*] n–space defined by the simultaneous vanishing of a number of homogenous polynomials[#]. If V and W are projective varieties, we can make the set $V \times W$ into a variety — this is not obvious for projective varieties, and relies on the fact that we can embed $P^n \times P^m$ in $P^{(n+1)(m+1)-1}$ (see Shafarevich, 1972, p.42). A function from one variety to another is said to be *regular* if in a neighbourhood of every point, it can be expressed by means of a homogenous rational function of degree 0.

A variety V is said to be an *algebraic group* if there is a group law on the points of the variety such that $f: V \rightarrow V$ with $f(x) = x^{-1}$ and $g: V \times V \rightarrow V$ with $g(x,y) = xy$ are regular mappings. A projective irreducible algebraic group is termed an *Abelian variety*. An abelian variety is always commutative (Shafarevich, 1972, p.153 Theorem 3). By Theorem 3 above, an elliptic curve with a distinguished point is an Abelian variety, since the formulae quoted in that Theorem are regular.

More generally, for any algebraic curve, we can construct the group of linearly inequivalent divisors of degree 0. It is possible to make this into a variety, and it then becomes an Abelian variety, known as the *Jacobian* of the curve, written as Jac(C) for the curve C. Theorem 3 can then be restated as saying that an elliptic curve is birationally equivalent to its own Jacobian and for most purposes we can identify the two.

---

[*] Projective n-space is defined as $\{(a_0,...,a_n)\ a_i$ in K, not all a_i simultaneously $0\}$ under the equivalence relation $(a_0,...,a_n) = (b_0,...,b_n)$ iff there is a c in K such that $a_i = b_i c$ for all i. We will write projective n-space as P^n where convenient.

[#] Note that we do not require that there be $n-1$ polynomials, or that the polynomials be independent.

Gauss-Manin Operators

Introduction

This chapter is devoted to the case of integrands which contain a transcendental parameter apart from the variable of integration, so that we can consider our problem to be the integration of a function in $\{K(x,y) \mid F(u,x,y) = 0\}$, where K is an algebraic extension of $k(u)$ for some field k and u transcendental over it. We shall use this notation, with u being the independent transcendental, and we shall use the prefix operator D to denote differentiation[#] with respect to u^*, and the suffix ' to denote differentiation with respect to x. This case is often more tractable than the case when there is no such transcendental, for integration with respect to x and differentiation with respect to u commute, so that if $G(u,x,y)$ is integrable, then so is $DG(u,x,y)$, $D^2G(u,x,y)$ and so on.

In this case we can sometimes determine that a divisor is not of finite order, so that we then know that the function is unintegrable. If we cannot do that, then we can

---

[#] We are going to differentiate with respect to u even if u does represent a constant transcendental parameter (such as e or π). We can do this because, since the parameter is transcendental, it cannot satisfy any algebraic equations relating to other constants in the integrand, and therefore its precise value does not matter.

This brings out the point made under "Theoretical Limitations" in Chapter 1 that we must know all the dependencies among our constants. For example, since it is not known whether e and π are algebraically independent, we cannot consider an integral involving both of them, since we do not know how to express dF/de in terms of $\partial F/\partial e$ and $\partial f/\partial \pi$. We could, of course, assume that they were independent and produce a result of the form "if e and π are algebraically independent, then F has no elementary integral".

[*] By this we will mean a total differentiation, taking into account the dependence of y on u caused by the functional relationship $F(u,x,y)=0$ (assuming that u appears effectively in this function). We have that $dy/du = -\partial F/\partial u/\partial F/\partial y$, so that $DG = \partial G/\partial u + (dy/du)\partial G/\partial y$.

determine a value u_0 in k for the parameter u such that the divisor $P(u)$ on $F(u,x,y) = 0$ is of finite order iff the divisor $P(u_0)$ is of finite order on $F(u_0,x,y) = 0$. In other words, we can reduce the problem to one not involving u. Recursively, we can reduce the problem to one with no transcendental parameters. We shall not discuss in this chapter how to solve such problems, rather they are left to the next two chapters. If $P(u_0)$ is of infinite order, then we know that $P(u)$ is too, and the problem is solved. Otherwise, let n be the order of $P(u_0)$ and consider $nP(u)$, $2nP(u)$ in turn until we discover what the order of $P(u)$ is, and we know that this process must terminate, though we have no idea when. In the case of curves of genus 1, we can use Cayley's determinant test in order to decrease the amount of work, as explained towards the end of the last chapter.

This work is based on two papers (1958, 1963) of the Russian mathematician Ju.I. Manin, and the reader is referred to them for the full generality of the exposition and most of the proofs, which we will just state. These papers do not make easy reading however, and both the English and Russian versions contain many misprints. Furthermore Manin's exposition is complicated by his desire to work with n parameters (and hence n differentiations) rather than just 1. We may, of course, have several parameters $u_1,...,u_n$, but we shall eliminate then one-by-one, using the methods of this chapter recursively, rather than try to consider them all simultaneously. I have been unable to conduct any experiments with curves with two parameters, so I have no firm ideas as to which or the two approaches (parallel or recursive) is better, but intuitively it seems that we want to make the problem as small as possible, by eliminating parameters, as soon as possible, rather than work with then all until we reach the end.

Example

Before giving the general theory, we will consider a worked example of this theory, taken essentially from Manin (1963, pp. 190-192). Consider the "general" elliptic curve $y^2 = x(x-1)(x-u)$ and take the ground field K to be a finite extension of $k(u)$ for some field k. Let $w = y^{-1}dx$ be a differential of the first kind (and since the curve is elliptic, all others are constant (in the sense of not depending on x or y) multiples of this form). Then if C is any closed curve on the surface $\{K(x,y) \mid F(u,x,y) = 0\}$, then $e = \int_C w$ is infinitely many valued (and an analytic function of u), and in fact the space of these is generated by an arbitrary pair e_1, e_2 of such periods[*]. These functions are the solutions of the Gauss linear differential equation $4u(u-1)D^2 e - 4(1-2u)De + e = 0$.

On the other hand, functions of the form $\int_O^P w$, where O is the point at infinity, are extremely important for the investigation of the geometry of the curve. Such functions are only defined up to some period, because we could choose any path to get from O to P, and satisfy the equation $I(P) + I(Q) = I(P + Q)$ modulo such a period, where $P+Q$ is the sum according to the group law (see the previous chapter). We can remove this ambiguity by acting on both sides of the equation with the Gauss operator, which destroys such periods, thus obtaining the function $J(P) = (4u(u-1)D^2 - 4u(1-2u)D + 1)I(P)$, which can be identified with an element of $\{K(x,y) \mid F(u,x,y) = 0\}$. Because of the linearity of the Gauss operator and the relationship between $I(P)$, $I(Q)$ and $I(P+Q)$, J defines a homomorphism from the group of points on the curve to the additive group of this function field.

This can be made explicit in the following way: first we observe that

$$[4u(u-1)D^2 - 4(1-2u)D + 1]y^{-1}dx = -2d(y(x-u)^{-2}).$$

If we then integrate round a closed curve, the right hand side is the integral of an exact differential, so vanishes, and on the left hand side the integration and the differentiation with respect to u commute, so that we have

$$[4u(u-1)D^2 - 4(1-2u)D + 1]\int_c y^{-1}dx = [4u(u-1)D^2 - 4(1-2u)D + 1]e(t)$$

thus giving the Gauss differential equation. Then

$$J(P) = [4u(u-1)D^2 - 4(1-2u)D + 1]\int_O^P \frac{dx}{y},$$

---

[*] A *period* is defined to be the integral of a differential of the first kind round a closed curve on the Riemann surface of the curve.

but this time we cannot commute the integration and D, because P depends on u. Write P as the point $(X(u),Y(u))$. Then if G is an algebraic function of x and u, rational in x and y, we can state that

$$D \int_O^P G(x,u)\mathrm{d}x = (DX(u))G(X(u),u) + \int_O^P DG(x,u).$$

We can then differentiate again to get that

$$D^2 \int_O^P G(x,u)\mathrm{d}x =$$

$$(D^2X(u))G(X,u) + (DX(u))[(DG(x(u),u)) + \partial G(X(u),u)/\partial u] + \int_O^P D^2 G(x,u).$$

In the special case under consideration, this gives us that[#]

$$J(P) = -2Y(X-u)^{-2} + D\frac{2u(u-1)(DX)}{y} + 2u(u-1)D(XDY).$$

Picard-Fuchs Equations

Let C be a curve of genus g defined by $F(x,y)=0$, where the coefficients of F lie in K, an algebraic extension of $k(u)$, and involve u effectively. Then we can find $2g$ closed curves on $C(u)$ such that any closed curve on $C(u)$ is continuously deformable to a sum of these $2g$ curves[*]. Let these curves be $c_1,...,c_{2g}$ and let $w_1,...,w_g$ be g linearly independent

---

[#] Observe the difference between the two halves of the second term in this summation. The first contains a total derivative of $G(X(u),u)$ with respect to u, while the other contains only a partial derivative, and is in fact the result of substituting $X(u)$ for x in $DG(x,u)$. The difference between these two is too subtle for many algebra systems, and this adds to the difficulty of implementing this work in a straight-forward fashion (see item 2 in Appendix 1).

[*] The precise wording is that we can find $2g$ curves $c_1,...,c_{2g}$ which form a basis for the 1-dimensional homology of the Riemann surface.

differentials of the first kind on $C(u)$. Define $L_{a,b}$ to be $\int_{c_b} w_a$, so that this is a period of the curve.

Lemma 1 For any a, the periods $L_{a,b}$ satisfy a linear differential equation of order $2g$ (or possibly less in degenerate cases):

$$p_{a,2g}(u)D^{2g}L_{a,b} + \dots + p_{a,1}DL_{a,b} + p_{a,0}L_{a,b} = 0$$

(i.e. the equation does not depend on b).

Proof: If w is a differential with no residues (i.e. of the second kind), then Dw is a differential with no residues (Manin, 1963, Corollary 2, p. 198). Then $w_a, Dw_a, \dots, D^{2g}w_a$ are $2g+1$ differentials of the second kind, so there is a linear relationship between them:

$$p_{a,2g}D^{2g}w_a + \; . \; + p_{a,1}Dw_a + p_{a,0}w_a = \mathrm{d}f$$

for some function f. Then we merely integrate round the periods, and the integral of $\mathrm{d}f$ round any closed curve is 0.

After Manin, we will call such equations *Picard-Fuchs equations*. The differential operator $L = \Sigma \, p_{a,i}D^i$ is termed a *Gauss-Manin operator*. We can endow the space of such equations of the form $L\int_c w = o$ with the structure of a module by allowing the more general equation* $\Sigma \, L_i\int_c w_i = 0$. Now if D is the divisor $\Sigma \, n_iP_i$, where the n_i are integers, define $\int^D f(x)\mathrm{d}x = \Sigma \, n_i\int_O^{P_i}f(x)\mathrm{d}x$, where the lower limit of integration is some fixed point O. This is clearly independent of the precise choice of O for divisors of degree 0.

If J is such a Picard-Fuchs equation, then we have that $\Sigma \, L_iw_i$ is an exact differential, say $\mathrm{d}z$. Then, if P and Q are any two points on an Abelian variety A, we can define#

---

* More formally (Manin, 1963, p. 199) a *Picard-Fuchs equation* is any relation of the form J: $\Sigma \, L_i\int_c w_i = 0$ where the w_i are differentials of the first kind and L_i are linear differential operators. Such equations can be defined on any Abelian variety A, but we will only be concerned with the case $A = \mathrm{Jac}(C)$.

We are using J in two senses, both for the Picard-Fuchs equation and for the operator which takes (P,Q) onto $z(Q) - z(P)$ (the so-called *Picard-Fuchs operator*), but this should cause no confusion.

$J(P,Q)$ to be $z(Q)-z(P)$, and this is well-defined (Manin, 1963, p. 202, Theorem 1) and is in fact $\Sigma L_i \int_Q^P w_i$, and it does not matter along which contour we integrate since the integral of an exact differential round any closed curve (i.e. the difference of two contours) is 0.

Furthermore $J(P,Q) = J(P,R) + J(R,Q)$, since we can choose a contour from P to Q which passes through R, and then split the integral at R (This is part (b) of Manin, 1963, p. 202, Theorem 1).

Picard-Fuchs operators as homomorphisms

We now make a fundamental remark about the relationship of the Picard-Fuchs operator J to addition of points on our Abelian variety:

$$J(P + R, Q + R) = J(P,Q).$$

The proof of this follows from the fact that differentials of the first kind are invariant under translations of our variety: this is made formal by Manin (1963, p.207, Lemma 12).

We now define $J(P)$ to be $J(O,P)$ where O is the zero of the group law on our Abelian variety. This brings us in line with the notation used in our example earlier.

Lemma 2 J is a homomorphism from the points of an Abelian variety (as an additive group) into the ground field.

Proof: In order to prove this, it is sufficient to prove that $J(P + Q) = J(P) + J(Q)$, since $J(O) = 0$ and O is the zero of the additive group on the Abelian variety.
$J(P + Q) = J(O, P + Q)$
$\qquad = J(O,P) + J(P, P + Q) = J(O,P) + J(O,Q)$ by the remark above
$\qquad = J(P) + J(Q)$.

Corollary 3 For any Picard-Fuchs equation J, the set of points of A of finite order lie in the kernel of the corresponding Picard-Fuchs operator J.

Proof: Because the ground field is torsion free and homomorphisms map torsion parts into torsion parts.

This means that the points of A of finite order lie in the intersection of the kernels of all the Picard-Fuchs operators. It would be wonderful if the converse were true, but that cannot be the case. To see this, let C be a curve over Q with a point P of infinite order. Now consider C and P over $Q(u)$. Then $P(u)$ is certainly still of infinite order, but we cannot say that it does not lie in the kernel of J, for J must take it and O to the same value, for both depend equally (not at all) on u. However, the next best thing is true, i.e. that this is the only way in which things can go wrong.

In order to explain this more precisely, we need a little more notation. This is taken from Lang(1959, p. 213), though we gain some simplicity by only considering the characteristic 0 case. Let K be any overfield of $\bar{k}$ for the purpose of this paragraph (though the application will be to K an algebraic extension of $\bar{k}(u)$). Let A be a variety over K. A pair (A',τ) is said to be a K/k-*trace* of A if A' is an abelian variety over k and τ is a homomorphism from A' to A and has finite kernel, such that, for any abelian variety B defined over k, and homomorphism $\beta:B - > A$ defined over K, there is a homomorphism $\beta':B - > A'$ defined over k such that $\tau\beta' = \beta$. This may appear a somewhat abstract definition, as indeed it is, but it defines the trace as that portion of A which is essentially independent of K/k (u in our case). The presence of τ is, in some sense, technical — the problem is that A' may have a few more points defined in it than we would like, which correspond to points defined over algebraic extensions of K.

Theorem 4 * If P is a point of A, and $J(P)=0$ for all Picard-Fuchs operators J on A, then there is an integer n such that nP (in the sense of the addition on A), lies on the image of the $k(u)/k$-trace of A. Furthermore if such an n exists, then $J(P)=0$ for all Picard-Fuchs operators J.

What this theorem means for our purposes is that, if $J(P) = 0$ for all J, then P is essentially independent of u. Now, there are infinitely many such Picard-Fuchs operators

---

* (Manin, 1963, p. 208, Theorem 2) I would like to thank Professor M.F. Singer for his invaluable assistance with this theorem.

V, but we need only consider a basis for the space of Picard-Fuchs operators, rather than all Picard-Fuchs operators. The dimension of such a basis is equal to the genus of the algebraic curve C in the case $A = \text{Jac}(C)$.

Divisors of Finite Order

Now let D be a divisor on an algebraic curve C. D corresponds to a point D' on $\text{Jac}(C)$, and D is rationally equivalent to 0 iff D' is of finite order on $\text{Jac}(C)$. If D' depends essentially on u, then it is certainly of infinite order, and if it does not, then the problem has been reduced to a simpler one. The way the reduction is performed is by substituting a value in k for u, so that the problem is reduced to one over k, rather than $k(u)$. Not every value of u will do - consider substituting $u = 0$ in $y^2 = ux^3 - 1$. The question of which values of u will work is called "Good Reduction", and is discussed in greater detail in chapter 8 (see the section "Criteria for Good Reduction, especially Theorem 8), where it plays a much more important part in the argument. It suffices here to say that there are only finitely many values of u which are not of good reduction (i.e which do not work) and there is a simple *a priori* test* to determine whether or not a value of u is of good reduction.

FIND__ORDER__MANIN

Input:

F(X,Y): the equation of an algebraic curve

> There is no actual need for this to be in primitive, rather than multivariate, form. The only explicit use made of F is in the calculation of the differentials of the first kind.

---

* The value u_0 in k of u is of good reduction if $F(u_0,x,y)$ is absolutely irreducible over k, and has the same genus as $F(u,x,y)$.

D: a divisor on the curve, written as $\sum_{j=1}^{M} n_j P_j$.
          We will sometimes write P_j as (X_j, Y_j).

U: a parameter over which the curve is defined.

Output:

INFINITE or an integer N, depending on whether D is

            of infinite or finite order. The integer N signifies that the image was of order

            N, so that we should consider ND, $2ND$, ... in our search to find the order of

            D.

[1] DIFF__1 :=

A linearly independent basis for the differentials of first kind on curve $F(X,Y)=0$.
Let G be the length of DIFF__1, viz the genus of the curve $F(X,Y)=0$.

            This can be done by a simple variant of Coates' Algorithm: see Chapter 3
            for details.

[2] For each W in DIFF__1 do:

[2.1] Let $A_{2G-1}, ..., A_0$ be indeterminates, and let A_{2G} be $1 - \sum_{i=0}^{2G-1} A_i$, so that the sum of all the
            A_i is 1 (because Picard-Fuchs operators are indeterminate up to constant multi-
            ples).

[2.2] Solve the equation $\sum A_i d^i W / dU^i = dR(X,Y)/dX$ for the unknowns A_i and the
            rational function $R(X,Y)$.

            The denominator of $R(X,Y)$ can be chosen to be the least common multiple
            of the denominators on the left hand side, and after clearing denominators
            the equation breaks up into a series of linear equations in the A_i and the
            coefficients of $R(X,Y)$, all of which can depend on U, but not on X or Y.
            Furthermore the degree of R is at most one more than the highest degree
            on the left hand side.

            Our Gauss-Manin operator J corresponding to the differential of the first kind W is now $\Sigma\, A_i \mathrm{d}^i/\mathrm{d}U^i$ and $J\,W = \mathrm{d}R(X,Y)/\mathrm{d}X$.

We now have to compute $\dfrac{\mathrm{d}^i}{\mathrm{d}U^i}\int_O^P W\mathrm{d}X$ for $1 \le i \le 2G$. This can be written as $\int_O^P \mathrm{d}^i W/\mathrm{d}U^i + B_i(X,U)$, where B is the contribution of all the other terms from the repeated differentiation.

[2.3] SUM := 0.

This will be used to accumulate $J(D)$ in.

[2.4] For j = 1 ... M do:

[2.4.1] B_0 := 0.

[2.4.2] For i = 1 ... 2G do:
$$B_i = \frac{\mathrm{d}B_{i-1}}{\mathrm{d}U} + \frac{\mathrm{d}X_j}{\mathrm{d}U}\frac{\mathrm{d}^{i-1}W}{\mathrm{d}U^{i-1}}) \mid_P \text{ (where } \mid P \text{ means evaluation at the } X,Y \text{ values}$$
of the point P_j).

The first term in this expression comes from the differentiation of the B term for the previous i, and the second term comes from the fact that $\dfrac{\mathrm{d}}{\mathrm{d}U}\int_O^P f(x)\mathrm{d}X = \int_O^P \dfrac{\mathrm{d}f(x)}{\mathrm{d}U}\mathrm{d}X + \dfrac{\mathrm{d}P}{\mathrm{d}U}f(p)$ (since $\mathrm{d}O/\mathrm{d}U = 0$).

$J(P)$ is now $\Sigma\, A_i \dfrac{\mathrm{d}^i}{\mathrm{d}U^i}\int_O^P W\mathrm{d}X$, which can be re-ordered as

$$\sum B_i(P,U)A_i + \int_O^P \sum A_i \frac{\mathrm{d}^i W}{\mathrm{d}U^i}\mathrm{d}x.$$

[2.4.3] SUM:=SUM + $n_j\Sigma\, B_i A_i + R(P_j)$.

The above expression is the previous formula multiplied by n_j and is the contribution of $n_j P_j$ to $J(D)$.

[2.5] If SUM is non-zero, then return INFINITE.

[3] For U0 = 0,1,2,...

   If GOOD__REDUCTION(U0,F,K)

Then do:

> This is not necessarily the best way of performing this choice of the value
> for U. For example, suppose the equation F depends on U and $\sqrt{U^2+1}$,
> and that 0 is not of good reduction. Then choosing U0 = 1 will give us a
> curve defined over $Q(\sqrt{2})$, whereas choosing U0 = 3 will give us one
> defined over Q. However appealing such intuitive choices may be, it is
> hard to devise a program for finding 'good' values of U in that sense.

[3.1] D:= Substitute U0 for U in D.

[3.2] F:= Substitute U0 for U in F.

[3.3] Return FIND__ORDER(F,D).

> FIND__ORDER is implemented as one of FIND__ORDER__MANIN (in the
> event that, even after substituting U0 for U, there is still a transcendental
> parameter), FINITE__ORDER__ELLIPTIC (see Chapter 7) or
> BOUND__TORSION (see Chapter 8).

Implementation

The implementation of this algorithm is technically fairly difficult, though few
mathematical problems are raised by it. As was mentioned earlier when discussing the
example, one major source of difficulty is the differentiation with respect to U and X
occurring in the same expression, and the need to distinguish between partial derivatives
with respect to U (as in step 2.2) and total derivatives (i.e. with the point P_j substituted
for X and Y, which may well depend on U). In order to do this I have found it easier to
write my own special-purpose top-level differentiation routines, rather than try to use
REDUCE's and manipulate REDUCE's data structure for derivatives, which has no
provision for distinguishing between total and partial derivatives. I do use a modified
version of several of REDUCE's differentiation routines for simpler parts of the task:
details of the modifications are given in Appendix 1.

Another major source of difficulty is the need for an efficient implementation, with as few calculations as possible being repeated. Since all the expressions involved are multivariate (with both the *Gauss-Manin parameter* and the variable of integration involved, "simple" operations such as the computation of greatest common divisors can be very time-consuming. This leads to the requirement for a variety of "look-aside" tables containing, for example, $d^i W/dU^i$ or the (partial) derivatives $\partial^i Y/\partial U^i$, which must be created and purged as appropriate. There are several other efficiency points: for example the linear equations in step 2.2 can be partially sparse[#], and it is necessary to take advantage of this in order to obtain an implementation with reasonable efficiency. Despite these and other tricks, this algorithm can still be very expensive because of the size of expressions that have to be manipulated, especially in step 2.2. When attempting to discover whether the divisor consisting of the point $(d(d-1),d(d-1)^3)$ with multiplicity 1 and the point at infinity with multiplicity -1 on Tate's curve (Appendix 2, Example 4) was of finite order or not (it is in fact of order 7) a carefully coded draft implementation of this algorithm consumed approximately 15 minutes CPU on the IBM 370/168 at IBM Thomas J. Watson Research Centre at Yorktown Heights[*].

Although the Gauss-Manin operator is generally of order $2G$ (where G is the genus of the curve), there are many special cases in which it is degenerate and has lower degree. For example, while the Gauss-Manin operator of a general elliptic curve has degree 2, it has degree 1 in the following special cases (Manin, 1958, p. 77): $Y^2 = X^3 + aX$ $Y^2 = X^3 + a$ $Y^2 = X^3 + a^2X + ba^3$ where b is any constant (i.e. not depending on X, Y or a) not equal to 27/4 (or else the curve ceases to be elliptic). Furthermore, in these cases

---

[#] More information on sparsity, and an example of how it can make equations much simpler to solve is given in Appendix 2 Example 4. There is also a general discussion of linear equations in Chapter 9.

[*] Unfortunately a more precise figure is not available because of a bug in the CMS-Batch simulation of OS installed at Yorktown Heights, which meant that the LISP timing features were inoperative and the CPU time had to be calculated by multiplying elapsed time by the "service ratio" (i.e. the ratio of CPU to elapsed time) produced by the operating system. However the figure is probably accurate to within 25%. The 370/168 is approximately 5% faster than the 370/165 installed at Cambridge, and on which the remainder of the timings quoted in this monograph were measured.

the Picard-Fuchs operators are extremely simple (being $\dfrac{XDa-2aDX}{2aY}$, $\dfrac{XDa-3aDX}{3aY}$, and $\dfrac{XDa-aDX}{aY}$ respectively). It is therefore important, in the interests of efficiency, to recognise these degenerate cases as early as possible, and this is not easy.

A further special case is the one where X_j is independent of U. In this case all the B_i are 0 and the computation simplifies considerably. This case frequently arises in practice, and hence has to be tested for.

Special Values of Parameters

In this section we shall suppose that our integrand $f(x,u)$ dx depends algebraically on u. This is not really a restriction, since if it depends transcendentally on u, we can replace a transcendental function of u by a new transcendental parameter u' without altering the problem, since we know that, if $f(x,u)$ is integrable, then its integral is defined over the *algebraic closure* of the original ground field, i.e. no new transcendentals can be introduced.

Proposition 5 If $f(x,u)$ depends algebraically on u, then the residues of $f(x,u)$ dx are algebraic functions of u (because they lie in the algebraic closure of the constant field).

If the algorithm FIND__ORDER__MANIN returns the answer INFINITE because one of the values of SUM was non-zero, one can conclude that the integral is not expressible in elementary terms. It might seem reasonable to consider those special values of u for which SUM happens to be zero, and ask whether the integral is elementary in this special case. This leads on the the more general question:

"*For what values of u is $\int f(x,u)$dx an elementary function?*".

Unfortunately, those values of u which make SUM zero are not the only values which can make the integral elementary, for there are several other ways in which the problem can reduce when a special value is substituted for u.

Let us now consider the various ways in which the substitution of a special value for the parameter u can alter the nature of the integration problem:

1) The curve can change genus. This can only happen for finitely many values of u, and we can decide what these values are by considering the canonical divisor, and the possibilities for it to degenerate.

2) The places at which residues of the integrand occur can change. There are only finitely many of these, and they can be detected by looking for all the values of u for which factors of the numerator and of the denominator coincide, or factors of the denominator coincide with each other (or for which the numerator or denominator change degree, to allow for the case of the factor $x - $ infinity).

3) The dimension of the space of residues can collapse. This is an exceptionally tricky case, and we will postpone a full discussion of it.

4) A divisor may be a torsion divisor for a particular value of u, even though not generally. This is where we started on this discussion, and these cases (of which there are only finitely many) can be detected by looking at all the roots (in u) of the functions SUM in the algorithm FIND__ORDER__MANIN.

5) The algebraic part may be integrable for a particular value of u, though not in general. These cases can be detected by looking at the equations generated in the algorithm FIND__ALGEBRAIC__PART to see when the contradicting equation, which proves that the function is unintegrable, becomes degenerate.

Thus we have shown that the number of "exceptional" values of the parameter u is finite (and these values are effectively computable) for cases 1,2,4 and 5.

Case 3 is substantially more difficult. As an example, we can have infinitely many values of u for which the Z-module of residues decreases in dimension:

Consider an integrand whose 4 residues are $1, -1, u, -u$; for example

$$\frac{1}{x\sqrt{x^2+1}} + \frac{1}{x\sqrt{x^2+u^2}}dx.$$

Then for every rational value of u the residue space becomes 1-dimensional, and hence potentially has to be considered as a special case.

Lemma 6 Let the Z-module of residues $(r_1,...,r_k)$ of $f(x,u)\,dx$ have dimension k. Suppose that there are values $u_1,...,u_k$ of u such that $f(x,u_i)dx$ has an elementary logarithmic part (without lying in cases 1,2 or 4 above) for $1 \leq i \leq k$ and such that the set of vectors $\{(r_i(u_a) \ 1 \leq i \leq k) \ 1 \leq a \leq k\}$ is of dimension k. Then $f(x,u) \ dx$ has an elementary logarithmic part.

Proof: There is an integer n such that the vector $(n,0,...,0)$ can be expressed as a linear combination, with integer coefficients, of the residue vectors $(r_i(u_a))$. Then the divisor d_1 corresponding to the residue r_1 must be a torsion divisor, because it has been expressed as the n-th root of a sum of torsion divisors. Similarly all the other divisors are torsion divisors, and hence the logarithmic part of the general integral must exist.

Theorem 7 If $f(x,u) \ dx$ is not elementarily integrable, then there are only finitely many values u_i of u for which $f(x,u_i)dx$ is elementarily integrable.

Proof: The only problem is case 3 above, for we have shown (and our arguments can easily be made completely rigorous) that there are only finitely many values which correspond to cases 1,2,4,5. So suppose that there are infinitely many values corresponding to case 3, but not to case 1,2 or 4. Then by Lemma 6 above, the Z-module spanned by the residue vectors $(r_i(u_a))$ is of dimension less than k, and so can be embedded in a space of dimension $k-1$. Then there is a linear relationship between $r_1(u),...,r_k(u)$ which is not true in general, but which is true for infinitely many particular values of u. but since the $r_1(u)$ are algebraic in u, by Proposition 5 above, this means that we have an algebraic expression which is not identically zero, but which has infinitely many roots, and this establishes the required contradiction.

Note that this Theorem is not completely constructive, in that we have shown no way of finding out what the finitely many values in case 3 are. That this problem is not completely trivial can be shown by the following example:

Let E be an elliptic curve over Q with a point of infinite order and a point of finite order (i.e. the infinite portion of the Mordell-Weil group is to have rank at least 1 and the torsion part is to be non-trivial, and such curves exist, as is shown by the tables in Birch & Swinnerton-Dyer (1963) or Swinnerton-Dyer (1974)) known as P and Q respectively. Let D be a divisor linearly equivalent to $3P$, and D' be a divisor linearly equivalent to $5P-Q$. Let $f(x)$ be a function on the elliptic with divisor of residues D, and $f'(x)$ one with divisor of residues D'. Then consider the function $f(x)+uf'(x)$, whose residue space is 2-dimensional for irrational u, and 1-dimensional for rational u. When u is irrational, the logarithmic part cannot be found, while if u is rational, say m/n, the divisor is $n(3P) + m(5P-Q)$, which is a torsion divisor only if the coefficient of P is zero, viz $u = -3/5$. This example demonstrates the necessity for the restriction that $f(x,u)$ should depend algebraically on u, because if we had written $f(x) + \sin u \, f'(x)$ then there would have been infinitely many solutions, viz. all the roots of $\sin^{-1} -3/5$.

Elliptic Integrals Concluded

The previous chapter (including the algorithm FIND__ORDER__MANIN) completely solved the problem of torsion divisors over ground fields containing a transcendental. We are therefore left with the case of ground fields all of whose elements are algebraic over the rationals, and this is the problem we will consider in this chapter (for elliptic curves) and the next. Furthermore, any particular definition of a curve and of a divisor can only involve a finite number of algebraics, so we can restrict our attention to fields which are generated from the rationals by extending with a finite number of algebraics, i.e. *algebraic number fields*. Before we can explore the torsion divisor problem over them, we first need to know more about their structure and possible computer representations, and this we discuss in the next section, amplifying the discussion of general algebraic expressions in Chapter 2.

Algebraic Number Fields

The theory of algebraic number fields is extremely extensive, and we shall only need a small portion of it. In general we will quote only the necessary results, and the reader is referred to Borevich and Shafarevich (1966) or Weiss (1963) for further details and proofs. There is some superficial confusion between the vocabulary of algebraic geometry and that of algebraic number theory, but it should be obvious from the context which meaning of a phrase is intended.

An algebraic number field can be written as $Q(l_1,...,l_k)$ where l_i is algebraic over $Q(l_1,...,l_{i-1})$. As explained in Chapter 2 (see also van der Waerden (1949), Vol I, pp.126-127) we can express this as $Q(l)$, where l is algebraic over Q, and in fact l is the root of a monic polynomial with integer coefficients (see Chapter 2), and furthermore $Q(l)$ is the same field as $Q[l]$. If K is an algebraic number field, we will denote by K_z the set $\{a$ in $K \mid a$ satisfies a monic equation with coefficients in $Z\}$, viz the set of *integers of K*. We observe that K_Z is a subring of K (van der Waerden (1949), vol II, p. 76), and therefore contains the ring $Z[l]$. The two are not necessarily equal, as is demonstrated by the case $l^2 = 5$, where $\frac{l-1}{2}$ is not in $Z[l]$ but satisfies the equation $x^2 + x - 1 = 0$. From our computational point of view, it is relatively hard to determine[*] whether or not an element of K lies in K_Z, and we will often substitute the more stringent test that it should lie in $Z[l]$ where this is legitimate.

K_Z is an integral domain, but not necessarily a unique factorisation domain[$]. The ideals of K_Z form a semi-group under multiplication, and we can associate the ideal $\{n*a \mid a \text{ in } K_Z\}$ with the element n of K_Z (call this the ideal (n)). The mapping $n => (n)$ is a semi-group homomorphism from K_Z (as a multiplicative semi-group) into the multiplicative semi-group of ideals, with kernel the units of K_Z (which map onto $(1) = K_Z$). If A and B are ideals of K_Z we shall write $A+B$ for $\{a+b \mid a \text{ in } A, b \text{ in } B\}$, which is also an ideal of K_Z. We may write (m,n) for the ideal $(m) + (n)$.

Theorem 1 (Weiss (1963) p. 132) The semi-group of ideals of K_Z has unique factorisation, in the sense that any ideal B can be written uniquely (up to order) as $\Pi P_i^{n_i}$, where the n_i are positive integers and the P_i are *prime ideals* of K_Z, i.e. they cannot be written as the product of non-trivial ideals.

Example. We can now consider the example in the footnote of non-unique factorisation in K_Z in the light of this Theorem. Let A be the ideal $(3,2+l)$ and B be the ideal $(3,2-l)$. Then $B^2 = (2-l)$ and $A^2 = (2 + l)$, and furthermore $A*B = (3)$. Hence $(9) = A^2B^2$ and in fact A and B are prime ideals of K_Z (see Weiss (1963) pp. 134-5).

Let p be a *rational prime* (i.e. a prime in Z in the conventional meaning of the word prime). Then p can be written as the product of prime ideals P_i of K_Z and the exponents

---

[*] An algorithm to do this is presented in the section on implementation at the end of the chapter.

[$] Consider the case $l^2 = -5$. Here we have $9 = 3 * 3 = (2+l)*(2-l)$ and none of 3, $2+l$, $2-l$ have any proper factors (Weiss, 1963, p. 134).

are termed the *ramification indices* of the prime ideals. We say that P_i is *ramified* if the corresponding exponent is greater than 1, and *unramified* if it precisely 1. p is said to be *unramified* (for the extension K of Q) if all the exponents are 1. We say that the P_i *lie over* p, and furthermore a prime ideal can only lie over 1 rational prime (consider norms, and the fact that two rational primes can have no common factor).

Theorem 2 (Dedekind. See Weiss, (1963), p.157) For any l (i.e. for any finite extension K of Q) there are only finitely many ramified primes.

If p' is a prime ideal of K_Z we can construct the quotient ring K_Z / p', where the definition of quotienting is that $a \neg b$ iff $a - b$ lies in p'. The relation $\neg$ is an equivalence, so the quotient is in fact a ring, known as the *residue class ring*.

Theorem 3 (Borevich & Shafarevich, 1966, p. 208, Theorem 2.) The residue class ring K_Z / p' is a finite field of characteristic p.

Elliptic Curves

There is a great deal of theory concerning torsion points on elliptic curves, some of which was mentioned in Chapter 5 as part of the general discussion of torsion. Unfortunately, despite the attention it has received from some of the greatest mathematicians, the subject is still very difficult, and many of the results involved are partial or non-constructive. As an example of this, we quote the following:

Theorem 4 (Manin, 1969) *For every algebraic number field K and prime p, there is an integer $n(K,p)$ such that, for all elliptic curves E defined over K, $| E_{\text{tors}}(K)^p | \leq p^{n(K,p)}$, where $E_{\text{tors}}(K)$ is the group of all torsion points of E defined over K, and the superscript p means the p-part of the group, i.e. those elements with order a power of p.

---

* There is an exposition of this subject by Serre (1971).

Unfortunately, except in the case $K=Q$, which was discussed in Chapter 5 under the heading "Mazur's result", we know very little about these numbers $n(K,p)$. About the only result is the following (Kenku, 1979):

Theorem 5 If K is a quadratic number field, then $n(K,2) \le 5$, where 5 is best possible.

In fact Kenku also shows that no point on an elliptic curve over an quadratic number field can have torsion 32.

This demonstrates the difficulties of finding general bounds for torsion on elliptic curves. Much more can be done if we consider a specific point, rather than trying to find bounds for all possible torsion points, and this we do in the next section.

Lutz-Nagell Theory

In this section we consider an elliptic curve $Y^2 = X^3 + aX + b$ where a and b are integers of an algebraic number field $K = Q[l]$. A point is said to be *integral* if both X and Y are integers of K (though if X is an integer of K, and Y lies in K, then it must be an integer of K).

Theorem 6 There are only finitely many integral points on any given elliptic curve.

Proof: This result was originally proved by Siegel (1929, see also Lang (1960) and Siegel (1969)), but this proof did not provide any information about the number of integral points or about their size. In many case it is now known (Baker, 1975, p.45) that all integral points satisfy $|X|, |Y| < \exp((10^6 H)^{1000000})$ where H is the maximum of the heights[$] of the coefficients of the curve.

Corollary 7 For fixed d, the number of points on any given elliptic curve with dX

[$] The *height* of an algebraic number is defined to be the maximum of the absolute values of the coefficients of its minimal polynomial. Therefore the height of a rational number is its absolute value.

integral, i.e. den(X) | d, is finite.

Proof: Consider the curve $Y^2 = X^3 + d^4aX + d^6b$, whose points are precisely (d^2X, d^3Y), where (X,Y) is a point on the original curve.

In fact we are only interested in a bound to the extent that it affects the computing time of the algorithm, and Baker's bounds are unduly* pessimistic.

Theorem 8 Let (X,Y) be a point of order precisely n on the elliptic curve $Y^2 = X^3 + aX^2 + bX + c$ defined over the algebraic number field K.

(1) If n is not a rational prime power, then X is an algebraic integer.

(2) If n is a power of the rational prime p, then pX is an algebraic integer.

Proof: The statement of the Theorem is almost precisely that of Lang (1978, p. 55, Theorem 2.1 as amplified by the remark attached to Theorem 2.2). The difference lies in case (2), where Lang states that the denominator of X divides the product of $P_i^{r(i)}$ where the P_i are prime ideals in K_z dividing p, and the $r(i)$ are rational integers $\leq e(i)/2$, where $e(i)$ is the ramification index of P_i as a divisor of p. But this product certainly divides $\Sigma\, P_i^{e_i} = p$, so the Theorem is proved.

We note that for case (2) to be non-trivial, the prime p must have ramification index at least 2 (else $r(i)$ would have to be 0), and this is very rare (in fact, if $a = 0$ we can replace 2 by 4, and this degree of ramification would be extremely unusual). We note also that we can consider a slightly more general form of the equation of the curve than the

---

* As an example of this, we can consider the elliptic curve 20A (using the notation of Swinnerton-Dyer (1975), Table 1). This curve can be expressed as $Y^2 = X^3 - 108X + 297$, and by Baker's bound has no integer points (and therefore no rational points of finite order) with $|X|$ or $|Y|$ greater than $E(e,10,10,6.92)$, where we use the notation that $E(a,b,c,...) = a^{E(b,c,d,...)}$ and $E(a,b) = a^b$ as a means of expressing nested exponents. In fact however, the calculations of Davenport and Stephens (using in part the Algorithm of Birch & Swinnerton-Dyer (1963)) show that the largest integral point on this curve is $X = 12$, $Y = \pm 27$, and this point is actually of finite order. As far as can be told, this is a "typical" example of the true size of integral points of finite order.

Weierstrass canonical form. This style of generalised Nagell-Lutz-Cassels Theorem is also discussed by Zimmer et al (1979).

Corollary 9 (Lutz-Nagell Theorem)[#] Any torsion point of an elliptic curve $Y^2 = X^3 + aX + b$ defined over the rationals, with integer a and b must have integral coefficients.

LUTZ__NAGELL [*]

Input:

F(X,Y): An elliptic curve in the form $Y^2 = X^3 + aX + b$, defined over an algebraic number field, with a and b integral over that field.

POINT: An integral point (X,Y) on that curve, also defined over an algebraic number field.

The two algebraic number fields do not have to be the same, since in fact we are not interested in what the algebraic number fields are as long as we can compute over them.

Output:

INFINITE if the point is of infinite order

N if the point is of order precisely N.

[1] Declare arrays XOLD[0:infinity], YOLD[0:infinity].

In fact these are probably best implemented as lists.

[2] XOLD[0] := X

[#] This result was stated in Chapter 5 as well, as part of the special machinery for elliptic curves over the rationals, but mathematically it belongs to this chapter.

[*] This algorithm was suggested by a procedure of Stephens (1970). I am very grateful to Dr. Stephens for this and many other discussions on computations on elliptic curves over the rationals.

YOLD[0] := Y

N := 0.

[3.1] If Denominator(X) not 1, then go to [4.1].

[3.2] For M:=0:N−1 do

      If X = XOLD[M] then do:

[3.2.1] If Y = YOLD[M] then N:=$(2^N)-2^M$

      else if Y = −YOLD[M] then N:=$(2^N)+2^M$

      else ERROR.

          We now know that POINT is of order dividing $2^N \pm 2^M$, but we would like a rather smaller number (in fact preferably the actual order) before we call DIVISOR__TO__FUNCTION. We can also assert that the factor of 2^M is necessary, for else we would have found a zero for smaller M and N.

[3.2.2] For each factor L (in order of increasing size) of N do:

      if POINT**$(L*2^M)$ = 0

      then return $L*2^M$.

[3.2.3] ERROR

[3.3] (X,Y) := 2* (X,Y) as addition of points on the curve.

[3.4] N := N+1

      XOLD[N] := X

      YOLD[N] := Y.

[3.5] If X = infinity

      then return 2^N.

[3.6] Goto [3.1].

          We can only go round this loop a finite number of times since we know that there are only a finite number of integer points.

[4.1] P := Denominator(X).

As explained in Chapter 2, our structure for algebraic entities places all the algebraic behaviour in the numerator, and leaves the denominator free, so P will be a rational integer.

[4.2] If P is not prime,

then return INFINITE.

[4.3] X:=XOLD[0]

Y:=YOLD[0]

N:=1.

We know now that, if POINT is of finite order at all, it has to be a power of P, so we consider expressions of this form directly.

[4.4] (X,Y) := P * (X,Y)

as addition of points on the curve.

N := P*N.

[4.5] If X = INFINITY

then return N.

The zero of the group law has X = Y = INFINITY.

[4.6] If denominator(X) | P

then goto [4.4].

[4.7] Return INFINITE.

Hardy's Statement

The simplest non-trivial case of the general problem of the integration of algebraic

functions in finite terms that of integrating functions defined on curves of genus 1. This
can be regarded as asking whether an apparently elliptic integral really is elliptic or not.
This problem was long thought to be insoluble, and Hardy summarised the classical position
when he said:

> "No method has been devised whereby we can always determine in a finite number
> of steps whether a given elliptic integral is pseudo-elliptic, and integrate it if it is,
> and there is reason to suppose that no such method can be given". (Hardy,1916
> pp. 47-8).

Note that previous work on elliptic integrals (Ng, 1974 and Carlson, 1965) did not
completely solve this problem, since they require an explicitly elliptic format (i.e. rational
function of x and the square root of a cubic or quartic in x) for the integrand. Conversion
into this format is done by our algorithm WEIERSTRASS__FORM, which relies heavily on
Coates' Algorithm.

The algorithms and techniques described so far in this monograph allow us to solve
this problem, i.e. to determine whether or not a given elliptic integral is pseudo-elliptic, and
to find the integral. Before we continue to consider curves of arbitrary genus (as we will in
the next chapter), we will explain how these algorithm fit together, since the algorithms for
curves of genus 1 are heavily specialised towards that case, and have also all been implemented, whereas some of the algorithms for higher genus curves described later have not.

Chapter 4 reduced the problem of integration to the problem of determining whether
or not a divisor was a torsion divisor, and finding its order if it was a torsion divisor.
Hence it is sufficient to describe a procedure that will do this, for elliptic curves over
arbitrary ground-fields (of characteristic 0).

FINITE__ORDER__ELLIPTIC

Input:

F(X,Y): The equation of a curve of genus 1.

This can be in any form, multivariate or primitive element, since it is converted to canonical form when required.

D: A divisor on the curve.

We assume that the divisor is not principal (i.e. of order 1), for we test for this before embarking on the FIND__ORDER procedure.

Output:

INFINITE if the divisor is not a torsion divisor,

else its order.

[1] If the constant field of F contains a transcendental parameter U, then do:

[1.1] If INFINITE = FIND__ORDER__MANIN(F,D,U)

then return INFINITE.

[1.2] For I = 2,3,... do :

If DIVISOR__TO__FUNCTION(F,I*D) is possible

then return I.

We could apply Cayley's test here (see chapter 5).

[2] If F(X,Y) and D contain no algebraic numbers,

then do:

[2.1] For I = 2,3,4,5,6,7,8,9,10,12 do:

If DIVISOR__TO__FUNCTION(I*D) is possible, then return I.

[2.2] Return INFINITE.

[3] If F is not in canonical form with integral coefficients,
    then do:

[3.1] F',D' = WEIERSTRASS__FORM(F,D).

[3.2] Write F'(X,Y) as $Y^2 = X^3+aX+b$.

[3.3] If a and b are not both integral
    then do:

[3.3.1] D:=lcm(denominator a, denominator b).

[3.3.2] b := $b*D^6$.

[3.3.3] a := $a*D^4$.

[3.3.4] For each (X,Y) in D' do:
        $X := X*D^2$;
        $Y := Y*D^3$.

[4] POINT := zero point of curve F', i.e. (INFINITY,INFINITY).

[5] For each P^N in D' do:
    POINT := POINT + N*P.
        This is done as addition of points on the Weierstrass canonical form F',
        according to the theory of Chapter 5.

[6] If Denominator(POINT)$ not 1, then do:

[6.1] A := A * $Denominator^4$.
    B := B * $Denominator^6$.

$ Interpreted as the least common multiple of the denominators of X and Y. We can in fact do better than this: see the section below on implementation.

We have multiplied the equation F', taken to be $Y^2 = X^3 + A*X + B$ by Denominator[6] and we now have to adjust the co-ordinates of POINT accordingly.

[6.2] Y := Y * Denominator[3].

   X := X * Denominator[2].

[7] Return LUTZ__NAGELL(F',POINT).

Implementation

The algorithms described in this section are fairly easy to implement given the algorithms for algebraic number fields. The major problem with working on elliptic curves is that one needs an expression for the zero point of the curve, which has x=y=infinity. The solution adopted for this problem is to invent a special kernel* "infinity", and to arrange that the addition routines for points on elliptic curves treat this kernel properly, which is not difficult in view of the following propositions:

Proposition 10 The only way in which infinity can occur in the representation of points on a canonicalised elliptic curve is in the representation of the point at infinity.

Proposition 11 The point at infinity is the zero of the group law on the curve.

Proposition 12 Two points on the curve can only add to give a point containing infinity if they have the same X co-ordinate and their Y co-ordinates add to give 0.

The algorithms described above are not optimal in several minor ways. The main reason for this is that we often, while clearing denominators from some expression or other, multiply throughout by the denominator raised to some power, when often a smaller expression would suffice. For example, when making the point (X,Y) integral, we multiply

---

* In the REDUCE-2 sense of the word: see Appendix 1 for details.

throughout by Denominator(POINT)[6], where in fact all we need is the least 6-th power greater than Denominator$(X)^2$ and Denominator$(Y)^3$.

The major problems lie in the underlying algorithms for algebraic number fields and algebraic integers. As was pointed out above, it is not always a trivial task to determine whether or not an algebraic expression is an algebraic integer. We present here an algorithm which decides this, and, more generally, determines the denominator of the algebraic number.

DENOMINATOR__ALGEBRAIC

Input:

  X: an algebraic number.

Output:

  N: the least integer such that N*X is an algebraic integer.

[1] If Denominator(X) = 1

        then return 1.

> This tests the denominator of X, regarded purely as an expression with no meaning attached to the algebraic expressions within it, and is purely a short-cut. It works because we know that all algebraic expressions generated by REDUCE (following the strategy of chapter 2) are algebraic integers.

[2] MATRIX[0]:=(1,0,.....,0).

> MATRIX will be a list of rows of a matrix, where row$_i$ ($i=0,1,...$) will be the coefficients of all the algebraic expressions in X^i.

[3] K:=0.

[4] While MATRIX is singular do:

[4.1] K:=K+1.

[4.2] MATRIX[K]:=Coefficients of algebraics in X^K.

This operation of taking the coefficients of all the algebraics in an expression must always yield a vector of rational numbers arranged in the same order, with the first item being the non-algebraic part.

[5] Let V[0] ... V[K] be such that V * MATRIX = 0 and V[K]=1.

Then the minimal monic equation with rational coefficients satisfied by X is
$$\sum_{i=0}^{i=K} V[i]X^i = 0.$$

[6] N:=1.

[7] For I:=0,1,...,K−1 do:

[7.1] For each prime P dividing V[I] do:

[7.1.1] For J:=0,1,...,K do:
$$V[J] := V[J]*P^{K-J}.$$

The effect of this is to replace X by X*P, and then to renormalise the V[J] to ensure that the equation is still monic.

[7.1.2] N := N*P.

[7.1.3] If P divides V[J]
then go to [7.1.1].

[8] Return N.

Curves Over Algebraic Number Fields

General Genus

The case of curves of arbitrary genus is much more difficult than the case of curves of genus 1, and there are no well-developed algorithms for this case. I have not been able to code any significant program to deal with this case because of the large number of subsidiary algorithms for which I do not have programs, though such programs have been written elsewhere, or can readily be written. Presented here, therefore, are the outlines of techniques which will enable one to bound the torsion of curves of arbitrary genus over algebraic number fields.

The matter is complicated by the fact that most of the necessary supporting mathematics is couched in very abstract language and much is contained only in the "folklore" of Algebraic Geometry. For this reason the results stated will be less detailed than in the rest of the monograph, and there will not always be complete references to support them. The theory to be described is true for curves of genus 1 as well as for curves of higher genus, and in fact lies behind much of the theory described in the previous chapter. We will therefore be able to illustrate[#] most of this work by examples over[*] curves of genus 1, and we will usually do this for simplicity although the main application of the work will be for curves of genus > 1.

The basic idea is similar to many processes in algebraic geometry: we reduce the problem to one over finite fields, in which we can compute explicitly and over which we

---

[#] There is an excellent illustration of these methods at work in Mazur & Swinnerton-Dyer (1974, p. 21, Lemma 1).

[*] Also there are many more computations and tables relating to elliptic curves (Swinnerton-Dyer, 1974, for example), so that it is easier to find suitable examples, and to explain their behaviour.

can perform complete searches, and then we can piece together the information from several finite fields in order to solve the original problem. In the terminology of computer algebra, we will adopt a *modular* approach.

Good Reduction

Let K be the common field of definition of the curve C and the divisor D, so that D corresponds to an element of the Jacobian of C as defined over K. Let p' be a prime ideal of K lying over the rational prime p. Let K' be the *residue class field* of K mod p', i.e. the field generated by the elements of the integers of K modulo the ideal p'. Let A be an Abelian variety over K (normally considered to be the Jacobian of C from our point of view, but the general theory does not require this). Let A' be the variety over K' defined by the same equations as A over K (i.e. A' is a *specialisation* of A in the sense of Mumford (1965)). We note that A' is defined over a finite field, and therefore has only a finite number of elements, which must all be torsion elements.

If p is any rational prime, and G is an Abelian group (normally a Jacobian divisor group but this is not necessary), then define the *p-part* of G to be those elements of G which have order a power of p. This is clearly a subgroup of G, and its order is a power of p. Define the *non-p-part* of G to be those elements of G whose (finite) order is coprime to p. This too is a subgroup of G. If every element of G is of finite order, then G is the direct product of its p-part and non-p-part for any prime p. Because of the manipulation of p-parts that we will engage in, we shall require the following algorithm:

MAX__POWER

Input:

   P: a positive integer, frequently a prime.

   N: a positive integer.

Output:

   Q: the largest integral power of $P \leq N$.

We will not bother to describe such a simple algorithm in detail.

    Following Serre and Tate (1968) we say that A *has good reduction*[*] at p' iff $A = A'$ x K, where the operation "x" means the taking of tensor products over the valuation ring of p'. The following result is essentially well-known:

    Theorem 1 If A has good reduction at p' which lies over p, then the non-p-part of the torsion subgroup of A is injected into the non-p-part of the torsion sub-group of A'.

    Corollary 2 The size of the non-p-part of the torsion group of A divides the size of A' (viewed as a group).

    We can also state the following result (from Serre and Tate (1968) or Shimura and Taniyama (1961)) as adapted to our circumstances and notation:

    Theorem 3 For a fixed K and A, there are only a finite number of primes of bad reduction.

    Theorem 2 of Chapter 7 implies that this is still true if we restrict ourselves to unramified primes, which we will do in the future. This means that we can use the following algorithm to reduce the general problem to that of torsion bounds modulo prime ideals.

---

[*] Roughly speaking, this condition means that A can be reconstructed from a knowledge of A' and K. This is a similar concept to that of *good evaluations* or *lucky primes* (see, e.g., Yun 1973).

BOUND__TORSION

(Version 1)

Input:

  K: an algebraic number field.

  F(X,Y): the equation of a curve defined over K.

Output:

  N: a bound for the torsion of F over K.

[1] For P = 2,3,5,... do:

  For each prime ideal P' of K with P' | P do:

    if GOOD__REDUCTION(F,K,P')

      then do:

[1.1] NON__P__PART := FINITE__BOUND(F,K,P').

  The algorithm FINITE__BOUND should give a bound for the torsion modulo the prime ideal P'.

[1.2] Go to [2].

[2] For Q = prime after P,... do:

  For each prime ideal Q' of K with

    Q' | Q do:

    if GOOD__REDUCTION(F,K,Q')

      then do:

[2.1] NON__Q__PART := FINITE__BOUND(F,K,Q').

[2.2] Go to [3]

[3] ANSWER1 := NON__Q__PART * MAX__POWER(Q,NON__P__PART).

ANSWER2 := NON__P__PART * MAX__POWER(P,NON__Q__PART).

Return minimum(ANSWER1,ANSWER2).

ANSWER1 splits the torsion group into its non-q-part and its q-part, while ANSWER2 does the converse.

This algorithm is not the only one which can use the mechanism we have developed to find a bound for the torsion of an elliptic curve. One possibility is to take 3 primes of good reduction rather than 2, and then observe that the p-torsion, for any p, will appear in the non-q-parts for 2 primes q, so that the torsion is bounded by the square root of the product of the 3 non-q-torsions.

Later in the Chapter we will see that something much better can be done when we know precisely what the torsion is over the finite field.

Torsion over Finite Fields

The aim of this section is to describe various ways of finding, or at least bounding, the size of Jacobian divisor group over a finite field, i.e. the implementation of the algorithm FINITE__BOUND. Much of this material comes from Lang (1959, chapter V, especially section 3)[#]. We require a piece of notation: if A is an Abelian variety defined over a field K, let $|A|_K$ be the number of points of A defined over K. Our first remark is that, if C is a curve of genus g, then Jac(C) is an Abelian variety of dimension g (see the discussion at the end of chapter 5). Hence we wish to discover $|\text{Jac}(C)|_K$ for the curve C under investigation. Unfortunately, this is not easily related to $|C|_K$. However we do have the following Lemma:

Lemma 4 (This is proved in Lang (1959, pp. 139-140).) Let A be an Abelian variety of dimension r over a finite field K with q elements. Let $\phi : A => A$ be the Frobenius

---

[#] I am grateful to Professor Sir Peter Swinnerton-Dyer for drawing my attention to this work, and for correcting many errors in my understanding of it.

endomorphism induced by the Frobenius* endomorphism on K. Then $|A|_K = \prod_{j=1}^{j=2r}(w_j-1)$ where the w_j are the characteristic roots of ϕ.

Lemma 5 (Lang (1959, p.138 Lemma 2 and Chapter IV section 3.) With the notation as above, $|w_j| = q^{1/2}$.

Theorem 6 $|\text{Jac}(C)|_K \leq (q^{1/2} + 1)^{2g}$.

Proof: From Lemma 4 and 5 above.

Criteria for Good Reduction

Theorem 7 If C has good reduction, then $\text{Jac}(C)$ does.

The converse is not necessarily true, but it does not seem worth trying to take advantage of those primes at which $\text{Jac}(C)$ has good reduction but C does not. I have managed to avoid any explicit construction of Jacobians in the programming, and I feel that the slight gains would be outweighed by additional complexity.

Clearly we will not have good reduction if the genus of C' is not the same as that of C, for then the Jacobians would have different dimensions. In fact the following is a necessary and sufficient criterion for good reduction to occur for C, and hence for $\text{Jac}(C)$:

Theorem 8 If C and C' have the same genus, and F' is absolutely irreducible (i.e. irreducible in all algebraic extensions of K'), then we have good reduction.

Let us consider the two halves of this test separately: the genus preservation part first. The genus of C' can never be more than that of C, so we have to detect those cases

---

* The *Frobenius endomorphism* is defined by $x = > x^q$. See Eichler (1966, p.249) for further details.

in which the genus decreases. This can happen in one of two ways: a differential of the first kind on C can cease to be a differential of the first kind on C', or the space of differentials of the first kind can contract. As an example of the first, consider $Y^2 = (X-1)(X + 1)(X + 2)$, which has genus 1 over Q with one differential of the first kind, viz. dX/Y. However, modulo 3 this function has a pole at $X=1=-2$, for $X-1$ is a local parameter there, and $1/Y$ behaves like $1/(X - 1)\sqrt{X+1}$. In fact, of course, the curve has genus 0 modulo 3. Since we know the differentials of the first kind as a result of computing the genus, this case is fairly easy to test for. The other possibility is that the differentials of the first kind will cease to be independent, and as an example of this consider the space curve $Y^2 = X^3 - 1$ and $Z^2 = X^3 + 2$. Here both dX/Y and dX/Z are independent differentials of the first kind over the rationals, but not when taken modulo 3, as one might expect.

Note that there is a possibility that we will reject some primes as not giving rise to good reductions when in fact they do, since we might inadvertently have an expression for a differential which was divisible by the prime in question. This is especially likely to happen over algebraic number fields, when we are dealing with prime ideals rather than straight primes, since in this case we cannot just divide out the prime. However, this can only happen finitely often, so we still have an infinite number of primes of good reduction available. This point may prove computationally embarrassing, but it does not affect the theory.

Example

We will consider the example of Tate's curve with $D=2$ from Appendix 2 example 4, and we will work over the rationals. This is not necessary, inasmuch as either Mazur's bound or the Lutz-Nagell approach will suffice, but it is a relatively easy case to work and explain. The equation is $y^2 = 4x^3 - 15x^2 + 8x + 16$. Clearly this does not have good reduction mod 2, because the equation reduces to $y^2 = x^2$ and not only is this clearly not irreducible but also the differential of the first kind, $1/y$, does not remain a differential of the first kind.

Modulo 3, the equation reduces to $y^2 = x^3 + 2x + 1$, which is irreducible and preserves the differentials of the first kind. Therefore the non-3-part of the torsion is at most $(3^{1/2} + 1)^2 = 7$ (after rounding down to an integer). The curve also has good reduction modulo 5, so that the non-5-part is at most $(5^{1/2} + 1)^2 = 10$ (after rounding down). Therefore the curve has torsion at most* minimum(7*9,10*5) = 50.

A Better Algorithm

In this section we will develop the consequences of knowing exactly how many points there are on the Jacobian of the curve over our finite field. Just as in the previous case we required the trivial algorithm MAX__POWER to extract p-parts, here we have an algorithm OCCURRENCES to do the same.

---

* In fact we can do rather better than this if we consider the various cases separately.

Case 1. The curve has 7-torsion. In this case the torsion group must have precisely 7 elements (this is, in fact, what happens).

Case 2. The curve has 3-torsion and 5-torsion. In this case the 3-torsion is bounded by 9, and the 5-torsion by 5, so the whole torsion is bounded by 45 (since introducing 2-torsion decreases the total). In fact this can be ruled out by considering reduction modulo 11, which gives 18 (and hence 15 once we have ensured that the group structure is maintained) as a bound for the non-11-part, and hence for the entire torsion.

Case 3. The curve has 3-torsion (but not 5- or 7-torsion). Then the maximum size of the torsion group is 10.

Case 4. The curve has 5-torsion (but not 3-torsion). In this case there can also be no 2-torsion (consider the non-3-part) so the torsion group must have order 5.

Case 5. The curve has only 2-torsion. In this case the torsion is bounded by 4.

Unfortunately I know of no good way of mechanising this sort of intuition, so we are left with the bound of

minimum(NON__Q__PART * MAX__POWER(Q,NON__P__PART),

            NON__P__PART * MAX__POWER(P,NON__Q__PART)).

OCCURRENCES

Input:

    P: a positive integer, frequently a prime.

    N: a positive integer.

Output:

    Q: the largest integral power of P dividing N.

We will not bother to describe such a simple algorithm in detail.

    Finding the number of points on the Jacobian of the curve is not easy, and I have no algorithm to suggest for doing it. In the case of curves of genus 1, then the curve is the Jacobian and there is no real problem (except that one has to be careful when counting multiple points, and to distinguish ordinary multiple points from ramified points). Nevertheless we present this algorithm, since it is clearly the correct way to approach the torsion divisor problem (from our current state of knowledge) in the case of algebraic number fields.

BOUND__TORSION

(Version 2)

Input:

    K: an algebraic number field.

    F(X,Y): the equation of a curve defined over K.

Output:

    N: a bound for the torsion of F over K,

                such that the true torsion is a factor of N (as opposed to version 1, where we merely knew that it was at least N).

[1] For P = 2,3,5,... do:

    For each prime ideal P' of K with

         P' | P do:

      if GOOD__REDUCTION(F,K,P')

        then do:

[1.1] NON__P__PART := FINITE__TORSION(F,K,P').

      The algorithm FINITE__TORSION should give the torsion modulo the

      prime ideal P'.

[1.2] Go to [2].

[2] For Q = prime after P,... do:

    For each prime ideal Q' of K with

        Q' | Q do:

      if GOOD__REDUCTION(F,K,Q')

        then do:

[2.1] NON__Q__PART := FINITE__TORSION(F,K,Q').

[2.2] Go to [3]

[3] ANSWER1 := OCCURRENCES(Q,NON__P__PART) * NON__Q__PART

$$\text{ANSWER1} := \frac{\text{OCCURRENCES(Q,NON\_\_P\_\_PART)} * \text{NON\_\_Q\_\_PART}}{\text{OCCURRENCES(Q,NON\_\_Q\_\_PART)}}$$

$$\text{ANSWER2} := \frac{\text{OCCURRENCES(P,NON\_\_Q\_\_PART)} * \text{NON\_\_P\_\_PART}}{\text{OCCURRENCES(P,NON\_\_P\_\_PART)}}$$

    Return gcd(ANSWER1,ANSWER2).

Furthermore this curve has 7 points on it, viz. $(0,\pm 1)$, $(1,\pm 1)$, $(2,\pm 1)$ and the point at infinity.

Computational Considerations

This section will describe some of the problems involved in the implementation of this "modular" algorithm, and outline some solutions. Since I do not have anything approaching a complete implementation of the work described in this chapter, this section may well not be complete,.

We know that the residue class field is finite of characteristic p, and it is in fact obtained from the integers of $Q(l)$ by identifying elements if their difference lies in the prime ideal. Since p lies in the prime ideal, we need only consider those elements of $Z[l]$ with integer coefficients between 0 and p, i.e. a finite set. Hence we can construct the residue class field by enumeration, though this may not be efficient in all cases. Since the field is of characteristic p, it contains a subfield isomorphic to the integers modulo p. The field is then an extension of the field of integers modulo p, and computation over such fields has been studied by Mignotte (1976).

We need to determine not only irreducibility but also absolute irreducibility in a residue class field in order to test for good reduction. We can factorise univariate polynomials in the subfield which corresponds to the integers modulo p by Berlekamp's Algorithm (Zimmer, 1972), but this is not sufficient. For example, in the residue field of 5 in $Q(\sqrt{2})$ (see the next section for a detailed discussion of this example) the polynomial $X^2 + 2 = 0$ factorises, whereas it does not factorise in the integers modulo 5. Hence, even for polynomials defined over the subfield, reduction to the subfield is not an adequate factorisation strategy.

Berlekamp (1970) presents an algorithm for reducing the problem of factoring univariate polynomials over a field of prime power order to that of factoring over the prime field, but this is not an easy process[*] and I have not yet implemented it. Alternatively, using the techniques of Trager (1976) generalised (where applicable) to finite fields and to multi-variate expressions of algebraic extensions (see Appendix 3 for details and algor-

---

[*] Mignotte (1976) describes the process as "complexe mais efficace".

ithms), we can reduce the problem to that of factoring a much larger polynomial over the field of p elements, which can then be solved readily. Once we have a univariate factorisation, we can grow it up to a multi-variate factorisation in almost[#] all cases, using essentially the same techniques as are used in p-adic factorising techniques (Wang, 1978 or Yun 1973 and 1976).

An Example over Algebraic Fields

Now let us consider the curve $Y^2 = X^3 + 8$, with differential $1/Y$. This does not have good reduction modulo 2 (since it becomes $Y^2 = X^3$, which is no longer of genus 1) or 3 (since it becomes $Y^2 = (X-1)^3$, also no longer of genus 1). It has good reduction modulo both 5 and 7, and the non-p-parts of the torsion are 6 and 12 respectively. Hence the torsion is at most 6 (in fact, the curve has no torsion over Q, but it does have one generator of infinite order (Birch & Swinnerton-Dyer, 1963), which has to map into a torsion divisor over a finite field).

Over the field $Q(\sqrt{2})$, the situation is slightly different. 2 and 3 are still primes of bad reduction (and, more generally, extending the ground field does not get rid of any primes of bad reduction). 5 is a prime in this field; and the residue class field has 25 elements, which we can represent by $\{(i + j\sqrt{2}), 0 \le i,j \le 4\}$ The curve has 36 points of finite order in this residue class field.

In $Q(\sqrt{2})$, the rational prime 7 splits into the product of 2 prime ideals, $< 7,4 + \sqrt{2} >$ and $< 7,3 + \sqrt{2} >$. The first of these is a prime of good reduction, and the residue class field has 7 elements, which we can equate with the numbers 0 to 6 modulo 7. The curve has 12 points of finite order over this field (exactly as in the case of Q, since the two residue class fields are isomorphic and the isomorphism preserves the curve). Thus

---

[#] it may be that all evaluations of the other variables in a multivariate factorisation are "unlucky" in the sense that the image ceases to be square-free. However, this can only happen for finitely many primes, so at the worst we can afford to treat this as a case of bad reduction and try a different prime.

with 36 for the non-5-torsion and 12 for the non-7-torsion, we are led to a bound of 12 for the torsion of the curve over the rationals as extended by the square root of 2. Although I know of no easy[$] way of discovering the torsion of an elliptic curve over fields other than the rationals, this curve does have a point of order 6 over the field $Q(\sqrt{2})$, viz $X = 4$, $Y = 6\sqrt{2}$, and furthermore twice this point (which is therefore of order 3) crops up in some comparatively simple integrals (see the footnote to Appendix 2 example 3).

Note that, though our point is of order 3 (or possibly 6) we have a computed bound of 12 by the second technique, but 252 by the first technique, so that this technique, while undoubtedly workable and effective in the mathematical sense of the word, has limitations for practical computation. One reason for this is that there are likely to be infinite order divisors as well, and these will tend to map into points of high (but finite) order in the torsion group corresponding to a good reduction, thus giving us unnecessarily high estimates.

Computationally, this is especially embarrassing because we only need the bound when there are divisors of infinite order, since if the divisor is of finite order we will find the order with or without a bound.

Conclusions

In this chapter we summarise the work described in this monograph, both the theoretical and the practical. We then describe some of the places where the theory has room for improvement, and places where the implementation is not complete or not as efficient as it could be. These latter fall into two classes, those related to the integration problem and those which are general problems of computer algebra. We then discuss the greatest open question: to what extent can these techniques be adapted to deal with transcendental functions as well as algebraic functions?

State of the Theory

In chapter 4 we have stated and proved Risch's Theorem (Risch, 1970), which reduces the problem of integration (for algebraic functions) to the torsion divisor problem. In chapter 6 we solve the torsion divisor problem for ground fields which are transcendental over the rationals, using the work of Manin (1957, 1963). In chapter 8 we use the technique of "good reduction" to solve the torsion divisor problem for ground fields algebraic over the rationals. Therefore we can solve the torsion divisor problem for all ground fields, and hence the integration problem for algebraic functions is completely solved.

The theory is therefore complete, although it is not always as nice as one would like. There are many areas where it could be more elegant, and some of these are described in the section "Further Theoretical Research" below.

State of the Implementation

While the theory is complete, the actual REDUCE-2 implementation is incomplete in several areas. Only one of these actually lies in the area of integration as described in this monograph, and that is the work of chapter 8, which has not been implemented in any significant amount. The implementation is also incomplete in that not all the sub-algorithms have been interfaced to the main system properly. I have had to develop many of them in isolation, either because the cost of running the main program to the point where the sub-algorithm is called was prohibitive or because of the difficulty in finding suitable test data.

The other major areas which require to be completed are:

1) The treatment of algebraics other than those which can be expressed in terms of square roots;

2) Sophisticated deduction of torsion bounds (see various comments in chapter 8);

3) Implementation of Cayley's test (see chapter 5);

4) Use of Lutz-Nagell as well as Mazur (see below and Appendix 2 example 7).

Those that are not discussed elsewhere are briefly mentioned below.

Algebraics. There are no great theoretical difficulties in the way of implementing general algebraic expressions. However, a substantial programming effort is required for the following reasons:

a) A polynomial no longer necessarily splits completely in an extension generated by one of its roots;

b) It is desirable (for efficiency reasons) to take account of generalisations of recent results (Trager, 1976 and 1978) about the defining field of an integral, and in particular to note that if the denominator of the integrand has an irreducible factor f(x), then the integrand has residues at all or none of the factors of f(x);

c) More generally, large savings can be made by adopting a more Galois-theoretic approach to problems.

The program is, in fact, integrated with a transcendental integration program (Norman & Moore, 1977), since they use the same factorisation code (see Appendix 1, item 5) and have the same interface to the user, viz. as an operator INT, whose first argument is the function to be integrated and whose second is the variable of integration. Together these programs amount to about 13,500 lines of REDUCE-2 source and about 285,000 bytes of compiled LISP (ffitch & Norman, 1977) object code (both figures being in addition to the underlying REDUCE-2 system). The program is now distributed for research purposes with REDUCE-2, and with the MTS operating system.

Lutz-Nagell. At the moment the procedure for elliptic curves (FIND__ORDER__ELLIPTIC described in chapter 7) does not use the Lutz-Nagell Theorem (see chapter 7)when the curve and divisor are elliptic and defined over the rationals, but instead uses Mazur's absolute bound of 12 (see chapter 5). The reason for this is historical, since the Mazur bound was implemented much earlier. In fact, as can be seen from Appendix 2 example 7, using the Lutz-Nagell theorem, which can tell us whether or not we have a torsion divisor, is much more efficient in the case when we do not have a torsion divisor, i.e. when we have logarithmic unintegrability.

Further Facilities Required

Many further facilities are required in order to make the program written so far into a complete usable system for the integration of algebraic functions. These include:

1) Transformation of complex logarithms into inverse trigonometric functions;

2) Efficient computation of greatest common divisors (see Appendix 1, item 3);

3) Factorisation (see Appendix 1, item 5);

4) Efficient computation of resultants;

5) More legible printing of complicated results including repeated lengthy kernels (see Appendix 1, item 1);

6) Rationalisation of linear equation techniques;

7) Heuristic transformations of the integrand;

8) Factorisation over algebraic number fields.

We shall describe some of these in more detail below.

Inverse Trigonometric Functions. Complex logarithms are unnatural to many people, especially as conventional integral tables nearly always present their answers in the form of inverse trigonometric and hyperbolic functions. It would therefore be nice if some of the answers from the logarithmic part could be expressed in these terms, although some are almost certainly best left as logarithms (e.g. Appendix 2 examples 4 and 5). One example where inverse hyperbolics are more usual is Appendix 2 example 1, where the integral of $1/SQRT(X^2-1)$ was given as $LOG(SQRT(X^2-1)+X)$, whereas a set of integral tables would give $cosh^{-1}(X)$.

Resultants. In the course of factorising over algebraic extensions of fields, we need to compute resultants (Trager, 1976, Algorithm SQFR__NORM; (see also under Algorithm SQFR__NORM in Appendix 3) see also chapter 2 and appendix 1 item 5). The algorithm I currently have is extremely crude, using the matrix determinant definition (van der Waerden, 1949, Vol 1, p. 84) of resultants and the algorithms (Lipson, 1969) present in REDUCE (Hearn, 1973) for computing determinants. There are much better algorithms available (Collins, 1971), as is brought out by Norman (1978b). Collins shows that two algorithms (both of them substantially better than ours) can differ by more than a factor of 10 in time consumed. This is an area of computer algebra in which substantial progress is still being made, and there is a new algorithm based on the work of Zippel (1979) which is probably worth investigating.

Linear Equations. The current program contains no less than 5[#] sets of routines for the solution of linear equations:

    a)   one   for   the   finding   of   linear   relationships   in   algorithms
    INTEGRAL__BASIS__REDUCTION and NORMAL__BASIS__REDUCTION;

[#] Furthermore the transcendental integration algorithm (Norman & Moore, 1977) contains 2 more, and a factorisation program normally incorporates one (Berlekamp, 1967). The REDUCE system (Hearn, 1973) itself contains one, based on Bareiss' method (Lipson, 1969), but it is unfortunately useless when algebraics have to be considered. As pointed out in chapter 2, we must reduce powers of algebraics while manipulating matrices and linear equations, but the method implemented has similar features to Euclid's algorithm, which, as shown in chapter 2, can collapse if algebraics are being reduced, because terms can be present in the input to a step but can disappear from the output when not expected to.

b) one for the linear equations needed in COATES to find the function which has the correct zeros;

c) one to express the residues in terms of a basis for the Z-module they generate;

d) one for use in the algorithm ALGEBRAIC__PART;

e) one to solve the linear equations required to determine the precise form of the Picard-Fuchs equation (see chapter 6).

These are all subtly different: for example (c) has to treat SQRT(A) as completely different from A, while the rest must treat SQRT(A)^2 as being the same as A. Furthermore (c) deals with integers, while the rest deal with rational functions or polynomials. An additional complication is that (a) must produce the first linear relationship, while the rest normally assume that they can re-order rows as they wish. Furthermore the equations in (e) are often upper triangular or nearly so, and hence probably require a special algorithm.

These linear equation problems can all be of widely differing sizes and have varying degrees of sparseness, and the routines must be capable of: dealing with small cases efficiently (else trivial problems take an inordinate time); dealing with sparse cases by sparse algorithms (else several problems will not be soluble in practice; see the footnote to Appendix 2 example 4); and being reasonably fast on medium-sized dense cases (large dense cases are probably impossible anyway). This probably requires some form of "meta-algorithm" that looks at the problem and then chooses the algorithm to use: straight-forward Gaussian elimination for the small case; a sparse-matrix method (probably Cramer's Rule and an algorithm for sparse determinants (Smit, 1976), with precautions to avoid evaluating the same minor more than once); and an adapted fraction free method (Lipson, 1969) when suitable, probably incorporating special code for dealing with sparse rows within the dense matrix (see the footnote # below).

This is not all that is required, though, for one would still require different copies of the code for polynomials, integers and rational functions, and for whether or not one simplifies powers of algebraics. In theory, since there are only a finite number of possibilities (at most 3x2, and not all are in fact required), the strategy of MODE-REDUCE (Hearn, 1976), which would involve compiling the same piece of source program 6 times, is sufficient, but I feel that a truly dynamic system, such as the proposed

SCRATCHPAD/370 system (Jenks, 1979) would be more efficient and would provide greater flexibility.

The code currently written for (b) above, even though it contains some simple improvements[#] on a standard Gaussian elimination to make it work better on sparse or partially sparse sets of equations, often takes a great deal of time to decide that a set of equations is insoluble. Since the insolubility of these equations is equivalent to the failure of Coates' algorithm, this can happen quite often. For this reason, I have implemented a modular check to ensure that the equations are soluble modulo a prime before trying to solve them over the true ground field. This should clearly be extended to all the linear equation solving systems, but the effort of implementing it 3 more times seems to outweigh the gains, while if one implementation were capable of solving all the linear equation problems then this quick check would be available to all.

This modular check is currently only used to give a "yes/no" answer, but more information could be gleaned from it. For example, using some of the ideas from the theory of probabilistic algorithms (Zippel, 1979) we can say that any zeros in the modular solution probably represent zeros in the real solution, and that before solving the general problem we ought to see if there is a solution in which all these elements are indeed zero. Unfortunately, if there is not such a solution we cannot assert that the original set is insoluble, but we have to try and solve this larger set. Nevertheless, the saving would probably be substantial.

Heuristic Transformations of the Integrand. There are several ways in which the input to the integration program can be transformed in order to make the program run faster. Perhaps the simplest is that described at the end of Appendix 2 example 6, where replacing A by B^2 makes the program run about 10% faster.

As a more sophisticated case, consider integrating $1/SQRT(X^2+1)$ +

The major one is a piece of look-ahead to choose the sparsest rows to eliminate next. In particular this means that rows with only one non-zero entry are automatically eliminated first, and thus cannot suffer from "fill-in" (Smit, 1976).

100/SQRT(X^2+10000). This has residues of 99,101,−99,−101 at the four places lying over infinity. The computation for cases like this is considered as example 8 in Appendix 2. If we tried to integrate one of the summands first, we can do this in 868 milliseconds, and the remainder of the integrand takes about the same time, thus giving a total time of under 2 seconds. We cannot merely divide integrands as we wish, though, because if $f(x)$ and $g(x)$ are unintegrable, it is still possible for $f+g$ to be integrable.

Therefore a good integration program would look for partial summands in the integrand, and see if they were integrable. If they were, then it would remove them and try to integrate the rest, having thus made the problem smaller.

Factorisation over Algebraic Number Fields. Our approach to factorisation has been based on Trager's (1976) work (see chapter 2) and there is, as far as I know, no suitable alternative algorithm for integrating over general algebraic extensions. When it comes to factorisation over algebraic number fields, there is a generalisation (Wang, 1976) of a scheme for factorising over the integers (Wang & Rothschild, 1975). See also Weinberger & Rothschild (1976). Where a suitable small prime exists, this is only slightly slower than factoring the same multivariate polynomial over the integers. If such a prime does not exist, then Wang's method is much less efficient, and it appears (Trager, 1976) that he has now implemented Trager's algorithm in this case.

For efficiency, therefore, we should use Wang's method where it is readily (see also under Algorithm ALG__FACTOR__2 in Appendix 3) applicable. The current factorisation scheme (see Appendix 1 item 5) is not suitable for this adaptation, since it is not based on the p-adic techniques of Wang & Rothschild (1975). The version under development by Norman & Moore (Norman, 1978b) is based on Wang's (1978) improvement to the algorithms of Wang & Rothschild, and it is not clear to me quite how one can adapt these improvements to the case of algebraic number fields. Further work on this would certainly speed up the integration process substantially in many important cases.

Implications for Algebra Systems

In this section we aim to collect all the points raised in this monograph which relate to the design and implementation of computer algebra systems, and make certain notes which should guide the implementor of any future algebra system. There are two main aspects of this: those points that are specific to integration, especially of algebraic functions, and those that I feel relate to the implementation of any large algorithm.

One general point about algebra systems that has emerged from the implementation of the integration system that is described here is that one needs "sophisticated" algorithms, such as p-adic techniques (Yun, 1976) or sparse matrix methods (Smit, 1976). It is possible to write a very successful system that is used by many mathematicians and scientists (e.g. REDUCE-2 (Hearn, 1973)) without the use of such techniques, but such a system is not adequate for our purposes (see Appendix 1 items 3 and 5 for examples). Furthermore these algorithms are, on the whole, genuinely long[$] and difficult.

One specific point that comes out is the question of algebraics. It has proved difficult to fit algebraics into REDUCE-2, and that seems to be (inasmuch as one can point to one specific cause) fundamentally because the system was not designed with them in mind. Therefore any future algebra system should, in my opinion, take account of the various problems posed by algebraics in the design stage, so that they fit into the overall structure of the system.

As has been mentioned before in this work (e.g. in the discussion of linear equations above), it appears that current computer algebra systems are insufficiently flexible,

---

[$] Although the counting of source lines is not a good measure of program complexity, it is hard to visualise a better measure. We therefore quote some figures as an illustration of this point. The REDUCE-2 system (Hearn, 1973), contains about 2000 lines of pre-processor and about 5500 lines of algebra system, of which about 1500 belong to the "high-energy physics package", which is irrelevant to our purpose. A draft version of a p-adic factorisation and greatest common divisor system (Norman, 1978b) is currently about 3500 lines, and is likely to be longer than this when it is finished.

inasmuch as the designer has produced a fixed list of types[#] and in order to add a new type, one must do all the work oneself. In particular the presence of a linear equation solver for rational functions does not mean that one can use it for polynomials, or for elements of finite fields. This leads to a great duplication of effort, and makes the implementation of a new algorithm using new data types much more tedious than it need be.

Two major attacks are now taking place on this problem: the **MODE-REDUCE** project (Hearn, 1976) and SCRATCHPAD/370 (Jenks, 1979; Jenks & Davenport, 1980). Both these approaches, which can be characterised very crudely as relying on static and dynamic mode analysis respectively, should lead to a great reduction in this problem. An important research project would be the implementation of an integration system on one of these systems, both in order to see if the promise of these systems is fulfilled and in order to act as a vehicle for future research in integration.

Future Theoretical Research

One major area for future research is Coates' algorithm. It is very often the case, especially when the divisor to be expressed has large orders at several places, that intermediate expressions in Coates' algorithm are much larger than the final answer. One particular form of this is that most of the functions in the normal integral basis (i.e. the list of functions with places no worse than those of the divisor) can have enormous expressions, but that the answer can be one of the smallest such functions. This seems to imply that Coates' algorithm is not necessarily the best way of finding a function with the correct zeros and poles, but I have no idea how to improve on it.

Closely connected with the previous topic is the question: "How long does Coates' algorithm take". I currently have no timing analysis for the algorithm, mainly because

---

[#] In the case of REDUCE-2 these could be listed as **ARRAY, MATRIX, STANDARD FORM** (i.e. polynomial), **STANDARD QUOTIENT** (i.e. rational function) and **PREFIX FORM** (which is the representation of unsimplified expressions).

there is no obvious way of predicting the number of reduction steps, either at finite places or at infinity, from the input.

Also on the subject of Coates' algorithm, there is the problem raised in chapter 3 of whether or not step 3 is necessary. The algorithm would clearly be both theoretically tidier and substantially faster if this step could be shown to be unnecessary under conditions which we could arrange always to fulfil. I˙ have made slight progress in this area, but there is a long way to go.

As appeared in chapter 8, there are few good techniques for working over residue class fields. It was suggested there that brute-force enumeration was an answer, but this is clearly an area where future research is needed.

As mentioned in chapter 5, the theory of integration has connections with the theory of continued fractions (Chebyshev, 1857 & 1860). There is a modern application of this (Schinzel, 1962), which can be used with a result of Mazur (1977) to give a completely different, albeit somewhat specialised, integration algorithm. There is clearly scope for further research on the relationship between these two areas.

The major area where the algorithm proposed in less than satisfactory is the question of curves of genus greater than 1 over algebraic number fields (including the rationals). In the case of curves over transcendental fields, we have Manin's work (see chapter 6) which allows us to decide whether or not a divisor is of finite order. In the case of curves of genus 1 over algebraic number fields, we can use the algorithm LUTZ__NAGELL (chapter 7) to determine if a divisor is of finite order, and what its order is. But in the remaining case, we have to compute a bound (often far larger than is necessary), and then test all divisors up to the bound. The situation would be much tidier if we had a mechanism for determining whether such a divisor was of finite order without such a search.

Extensions to Transcendental Functions

As mentioned in chapter 1, the problem of integrating purely transcendental functions was solved in a theoretical sense more than 10 years ago (Risch, 1969; see also Moses, 1967 & 1971; Norman & Moore, 1977; and the survey in Norman & Davenport, 1979). This monograph has shown how to solve the problem of integrating purely algebraic functions, and a large amount of the required code has actually been written.

This raises the question of mixed functions, i.e. those that are transcendental but not purely transcendental, e.g. $\sqrt{1 + \log x}$ or[*]

$$\log (\sqrt{a} + \sqrt{b-x})/\sqrt{(1-x^2))(b-x)(a-x)}$$

(Caviness, 1978, considered by Davenport (1979c)).

(Liouville's) Theorem 1 [$] Let F be a differential field with an algebraically closed constant field K. Let f lie in F, and let g be elementary over F with $g'=f$. Then there are $v_0, v_1, \ldots, v_k$ in F and $c_1, \ldots, c_k$ in K such that $f = v'_0 + \Sigma c_i \frac{v'_i}{v_i}$, i.e. $g = v_0 + \Sigma c_i \log v_i$.

We can call v_0 the *rational part* of the integral, and the rest will be termed the *logarithmic part* (the terminology is slightly different from that of chapter 4, where v_0 was termed the algebraic part: the reason is that v_0 is rational in the monomials occurring in f, which in chapter 4 were always algebraic, but which may now be transcendental as well).

---

[*] Note that it is not that easy to decide whether a function is of this form or not: One might think that $\sin x + \sqrt{1 - \sin^2 x}$ was not purely transcendental, but it can be written as $\sin x + \cos x$ and then expressed in terms of $\tan(x/2)$. This is the whole area of "Structure Theorems" (Epstein, 1979 or Rothstein & Caviness, 1979), and is a closely related area of research to that of integration.

[$] Stated by Risch (1970). The fundamental theorem on which this work is based (see chapter 4) is deduced by Risch as a corollary of this theorem.

If we know the logarithmic part, then the rational part can be found by extensions of the techniques of Risch (1969)[#] as implemented by Moses (1967, 1971) (see also Norman & Moore, 1977). As in chapter 4, therefore, the main problem is the logarithmic part.

The logarithmic part is determined by the residues of the integrand in precisely the same way, but the integrand is no longer defined over a curve, it is defined over an arbitrary variety; e.g. the integrand

$$\log (1 + \sqrt{1 + x^2}) + \frac{\sqrt{1 + \log x}}{\log x}$$

should be regarded as $Z + V/W$ in the field $\{K(X,Y,Z,W,V) \mid Y^2 = 1 + X^2, \ V^2 = 1 + W\}$. There is much less theory of divisors over arbitrary varieties than over curves, and I see no way to generalise the algorithms of this monograph to arbitrary varieties.

However, it is frequently the case that the variety can be regarded as the product of a curve and a hyperplane, and that the places at which the divisors are defined can be regarded as the product of points on the curve and the same hyperplane. In this case we need merely work on the curve, and the arguments of this monograph go through.

Example 1. $\log (\sqrt{a} + \sqrt{b-x})/\sqrt{(1-x^2)(b-x)(a-x)}$ (Caviness, 1978). Write $C^2 = A$, and $Y^2 = B - X$, so that the integrand is now $\log (c + y)/\sqrt{(1-(b-y^2)^2)(c^2 - b + y^2)}$. This integrand has no residues (this is not computationally trivial, but the mechanism of Puiseux expansions can be applied to the square root). Therefore the integrand is purely rational (in Y, the logarithm and the square root), and in fact is unintegrable.

[#] In the special case where the algebraic extension is a simple radical one, it would appear that the techniques of Trager (1979) extend gracefully to the case of mixed integrands. However, these techniques only address the algebraic part at the moment, and it is my belief that the logarithmic part is substantially more difficult.

Example 2. Consider the function

$$\frac{\log x(x^4 + yx^3 - x^2) + y(x^3 + x^2 - x) + x^4 + x^3 - 2x^2 - x + 1}{x(y + x)(x^2 - 1)}$$

where $y^2 = x^2 - 1$. This function has residues of ± 1 at the two hyperplanes lying over $x =$ infinity. Therefore this function has a logarithmic part of $\log(y+x)$ (as in Appendix 2 Example 1). The rational part is then $\log(x)*y$, and the function is integrable. In fact my program when running with the program of Norman & Moore (1977) is capable of integrating it.

Example 3. Consider the integrand

$$\frac{\log^2 x \, yx^2 + (\log x + 1)(yx^2 - y + x^4 - x^2)}{x(\log x + y)(x^2 - 1)},$$

where $y^2 = x^2 - 1$ as before. This integrand has residues of ± 1 on the two hyperplanes $y = \pm\log(x)$. We can then argue that the logarithmic part must be $\log(\log(x)+y)$, even though any attempt to do this computationally would probably rely on heuristics, e.g. replacing $\log(x)$ by $z = y+\log(x)$. We can then deduce that the algebraic part must be $y\log(x)$.

Example 4. The previous examples of mixed integrals have all been attackable by the strategy described above and the work of the rest of this monograph: this example I see no way of solving this way. Consider $\log(y + \log^2 x)$, where $y^2 = x^2 - 1$ as in the previous two examples. The derivative of this function is

$$\frac{2 \log x(x^2 - 1) + yx^2}{x(x^2 - 1)(\log^2 x - y)}:$$

let us consider the problem of integrating this. Intuitively a term of the form $\log(\log^2 x + y)$ might appear reasonable, but we are trying to produce algorithmic methods not heuristic ones. The integrand has a residue on the sub-variety $\log^2(x) = y$ of the variety $y^2 = x^2 - 1$. The residue is ± 1, depending on which portion of the variety we are on, and with special values at the points $x = \pm 1$. While it is possible to study this and assert that $\log^2(x) + y$ has the correct zeros and poles, I currently see no way of taking this step algorithmically.

Forms of Unintegrability

There are two distinct forms of unintegrability which can arise when trying to integrate algebraic functions: the first, which we shall call *logarithmic unintegrability,* occurs when we have a divisor of infinite order, so that the residues of the integrand cannot be accounted for by means of a logarithm in the integral; the second, or *algebraic unintegrability,* occurs when we can find a complete logarithmic part, but the function is still unintegrable.

There is a simple example of algebraic unintegrability in Appendix 2 example 2, where the integrand is in fact a differential of the first kind on its defining algebraic curve. In general, when a function is algebraically unintegrable, we can choose a function whose derivative has as many of the negative Puiseux expansion coefficients as possible (see algorithm FIND__ALGEBRAIC in chapter 4), and then the remainder is, in some sense, a minimal algebraic part. In the particular case of elliptic curves, there is an extensive theory (Whittaker & Watson, 1927) of these *elliptic integrals,* which we can take advantage of by transforming the curve into a standard form (algorithm WEIERSTRASS__FORM in chapter 5). Hence, in this particular case, we can express a non-elementary algebraic part in some useful standard form: it would be interesting to know how far this can be generalised, and what its relationship is to the work of Ng (1974).

There is an example of logarithmic unintegrability in Appendix 2 example 7. This integral is extremely interesting, since it behaves precisely like a logarithm (inasmuch as its derivative is a differential of the second kind) but yet is not one. There is clearly scope for further research on functions of this kind.

Summary

There now exists an algorithm for the integration of algebraic functions, and a program which implements a substantial part of it, in particular enough to disprove a conjecture of Hardy (see p. 100). There is some way to go before the rest is implemented, but the difficulties appear to be practical rather than theoretical. We have seen, in the previous two sections, that there is scope for extending the algorithm in the direction of transcendental functions, and some scope for extending our definition of "elementary" to include elliptic integrals.

Changes to REDUCE-2

This appendix describes the various changes made to REDUCE-2 (version of March 1978), which were needed to support the integration system described in this monograph. These are described here, not in order to draw attention to any shortcomings of REDUCE-2, but to highlight the difficulties that face any implementor of the integration algorithms described in this monograph. It is worth noting that an attempt by the present author to implement these algorithms under SCRATCHPAD (Griesmer et al., 1975) fell foul of much worse problems.

An important concept in REDUCE-2 is that of a *kernel* (Hearn, 1973, Chapter 5), which for our purposes can be regarded as a variable or a simplified* application of an operator (e.g. a structure like SQRT(2) or LOG(X)).

1 - Printing.

A change was made to the REDUCE-2 expression printing routines to print the structure (SQRT -1) as i rather than as SQRT(-1). This made many expressions significantly shorter, and rendered them easier to read as well. This is not necessary in standard REDUCE-2, because there I is treated as an independent variable which satisfies $I^2 = -1$ (see item 4 below).

Even with this change, many expressions were still taking several pages to print, much of which was repeated printing of the same complicated square root. In fact an early

---

* "Simplify" has a technical meaning in REDUCE-2, as in much of computer algebra. In REDUCE-2 *simplification* is a specific operation, whose aim is to convert expressions into a standard form, which should be canonical. In particular, each "operator" (e.g. SQRT, LOG) has associated with it a simplification routine whose task is to convert any applications of this operator into the standard form. Item 4 below discusses the SQRT simplifier in detail.

attempt to find the $X(7)$ function (see Appendix 2 Example 4) took more than 5 seconds cpu time to print an intermediate expression simpler than the final answer. A further modification was then applied to the REDUCE-2 expression printing routines which had the effect of replacing a SQRT(...) expression which was equal to the last such expression printed (as part of the same formula) by the symbol "lastsqrt". Thus $SQRT(A^2+X^2)$ Appendix 2 Example 4 contains a very powerful example of this.

In fact this is not a very good solution in the case where there are several square roots present in an expression: for example

$SQRT(A^2+X^2)$ *A + $SQRT(A^2+X^2)$ *X + SQRT(A)*A + SQRT(A) + $SQRT(A^2+X^2)$

would print as

$SQRT(A^2+X^2)$ *A + lastsqrt*X + SQRT(A)*A + lastsqrt + $SQRT(A^2+X^2)$.

A much better format (but one which is substantially harder to implement within the existing REDUCE structure)# is one which replaces each SQRT(...) expression which occurs more than once by a symbol of the form "ans1" or similar, and prints a list at the end of the expression, so that the example above would then print as:

ans1*A + ans1*X + ans2*A + ans2 + ans1

where ans1 = $SQRT(A^2+X^2)$

       ans2 = SQRT(A).

2 - Differentiation.

The REDUCE-2 differentiation routines maintain* an association list of kernels (other than atoms) and their derivatives, so as to ensure that the derivative of a kernel (which may be quite a complicated expression) is only computed once. The first disadvantage of this, as far as we are concerned, is that the integration algorithm requires quite a lot of

---

Nevertheless Professor Hearn has implemented this feature in the REDUCE version of 15th. April 1979 (the one after the version which is currently used in the integration system). I have been able to run some integrals under this new version of REDUCE, and an example of the output can be seen in Appendix 2 Example 5.

* I believe that this feature has been removed in the latest version of REDUCE-2 for essentially the reasons outlined below.

differentiation, often at different places, so that the same expression will be represented differently (e.g. as SQRT(X-2) at X=0, but SQRT(1-2X)/SQRT(X) at infinity) and therefore the savings are not as large as might be hoped, while a substantial amount of store (in one case, as much as 40Kbytes, which would have the effect, on the IBM 370/165 installed at Cambridge University, of increasing the charge for the job by 12%) is being used for storing the list. A much more serious effect is that the same answer is always returned for a differentiation of a kernel, even if the rules for the dependence of variables have changed in the interim. This can be extremely embarrassing in the computation of Gauss-Manin operators (as described in Chapter 6) because sometimes x depends on u and sometimes it does not. For these reasons the association list feature was removed from the REDUCE-2 differentiation package.

The Gauss-Manin operator package has a further requirement which requires changes to the REDUCE-2 differentiation package. If x is the variable of integration and u the "Manin parameter", then we sometimes wish to be able to calculate derivatives allowing for a dependency of x on u. With standard REDUCE-2, this is not possible if x is a variable (as opposed to a function). Allowing this possibility requires the introduction of an extra test into the REDUCE differentiation package, and this has been performed.

3 - Greatest Common Divisors.

The greatest common divisor package currently implemented in REDUCE-2 is a modified polynomial remainder sequence method (Hearn, 1979). While this is almost certainly the best such algorithm for a wide variety of problems, it can still take an inordinate amount of time to decide that two polynomials (especially multivariate ones) are coprime.

As has been remarked (Norman, 1978b), there are much more efficient ways of doing this, and I have followed Norman's first strategy of adding a modular check* for co-

---

* I am grateful to Dr. Norman for permission to use his program to do this, and for his assistance in implementing it.

primeness before entering REDUCE's g.c.d. routines. On the system available at Cambridge, computing modulo numbers less than 2^{24} can be performed particularly cheaply, and a prime this large is "unlikely" to be unlucky (See Zippel, 1979). I have not implemented Norman's second technique (a fully p-adic greatest common divisor, i.e. an EZGCD (Yun, 1973)) since there is currently# no multivariate implementation available in REDUCE-2. Even the first technique ensures that calculations which are not complete in 30 seconds in standard REDUCE (and which show no prospect of finishing) can take less than 1 second.

4 - Algebraics.

The treatment of algebraics in standard REDUCE-2 is, as described in chapter 2, insufficient for our purpose, since we need canonical expressions at (almost) all times. The treatment of algebraic expressions which I have added is only complete for those generated by means of the SQRT operation (allowing nested square roots), but the principles extend to arbitrary algebraics.

In order to manipulate expressions involving algebraic quantities, there are (in the integration system) analogues of the standard REDUCE-2 functions for manipulating expressions (e.g. ADDSQ to add two rational functions, MULTF to multiply two polynomials) which ensure that the output is canonical (i.e. no algebraic occurs to a power equal to or greater than its degree and there are no algebraics in the denominator of any expressions). Logically these routines (whose names are formed by prefixing the standard REDUCE names with the character *) operate by calling the corresponding standard REDUCE function and then applying the functions SUBS2Q (or SUBS2F) and SQRT2TOP. SUBS2Q and SUBS2F are REDUCE-2 functions which apply substitution rules of the form:

FOR ALL X LET SQRT(X)**2 = X,

thus eliminating all occurrences of square roots raised to powers. SQRT2TOP is a locally written routine that ensures that the denominator of any expression does not contain any

---

However one is being written at Cambridge University (Norman, 1978b), and I hope to
  be able to acquire this in the fairly near future.

algebraic expressions, by multiplying the numerator and denominator by the conjugate of the denominator.

In fact the routines do not all work this way, but try to deal with powers of algebraics as they arise, rather than storing them and attempting to get rid of them later. It must be emphasised that this somewhat disreputable piece of code is equivalent to that described in the previous paragraph, and is only being used for efficiency reasons. Implementing algebraics this way rather than in the straight-forward way of the previous chapter can often save up to 50% of the time of a run, and more in nested cases (Appendix 2 Example 6), where the straight-forward code has to operate several times on an expression to eliminate those powers of inner square roots that arise from the elimination of powers of outer square roots.

5 - Factorisation.

Factorisation of polynomials is important for integration: both as a prerequisite for the manipulation of algebraics (see item 6 below) and for determining the poles and zeros of a function. In the latter case our interpretation of factorisation is somewhat different from usual, since we must be prepared to extend the ground field if necessary in order to obtain all the zeros and poles. For example, x^2-2 does not factorise over the integers, but $1/(x^2-2)$ does have a pole at $\pm\sqrt{2}$.

The obvious solution to this is to take a standard program for factorisation of multivariate polynomials over the integers (Wang, 1978), add Trager's techniques for extension to factoring over algebraics (Trager, 1976; see also Chapter 2), and be prepared to extend the ground field by roots of any polynomial which does not factorise in the current ground field.

Unfortunately there is no factorisation package available in REDUCE-2. I was fortunately able to borrow a package developed by Dr. A.C. Norman and Mrs. P.M.A. Moore as part of their work on integration of transcendental functions (Norman & Moore, 1977)* and use it in the integration program. This package does not always factorise polynomials of degree greater than 4, but it will factorise all univariate polynomials. In the rare cases when I have needed factorisation of higher powers, I have either performed the factorisation by hand or have used the MACSYMA (Bogen et al, 1977) system to obtain the factorisation and then inserted it in the REDUCE-2 based integration program by hand.

In the near future I hope to be able to use a new factorisation package being developed by the same authors (Norman, 1978b) which will be capable of factorising all polynomials and which should be significantly faster in many cases than the current implementation.

6 - Uniqueness of Algebraics.

As explained in chapter 2, it is very important to ensure that new algebraic expressions are created only when there is no existing expression for the same value, otherwise we cease to have canonical expressions. For example, if SQRT(2) exists, we are not allowed also to create SQRT(8), since that can be written as 2*SQRT(2). Since the current implementation deals only with algebraics that can be expressed in terms of square roots, the code required to ensure this is written as a routine (SIMPSQRT) which ensures that any SQRT has the the required uniqueness.

The task is unfortunately complicated by the necessity to do this "locally" (i.e. at each place separately), as described in chapter 2. To do this, SIMPSQRT has to maintain a

---

* I am grateful to Dr. Norman and Mrs. Moore for permission to use their code, and for their assistance in interfacing it to my programs. I am also indebted to Dr. D. Dahm of Burroughs Corporation for several improvements to this package, and to Prof. A.C. Hearn for communicating them to me.

list of all the SQRTs defined over each place, and has to use the correct list each time it is called. The actual computation follows very much the line described in chapter 2 (some of the algorithms are presented in Appendix 3): Trager's (1976) algorithms followed by factorisation (see above). Since this can be extremely expensive[#] there is an option provided in the integration package to switch most of this checking off. Clearly, if the program does produce an integral without the checking, then the same expression (which may involve linearly dependent algebraic expressions) really is the integral; so that the full checking is only necessary when the program thinks that the function is unintegrable without it.

[#] Testing to see if a quadratic factors over a field containing 4 independent square roots involves factorising a polynomial of degree 32.

Examples

Here we collect several examples of the workings of the various algorithms described in this monograph. Although the techniques are presented in a fairly standard mathematical form, it should be emphasised that these examples have all been tackled by computer programs working essentially identically to the explanations given in this monograph.

Example 1: Simple Logarithmic.

Consider the problem of integrating $1/\sqrt{X^2-4}$. The integrand could have residues at ± 2 or at infinity. While the integrand (regarded as a function of X) has a pole at ± 2, when regarded as a differential it does not, since $X - 2$ is not a local parameter at $X=2$, and we have to express the differential in terms of $\sqrt{X-2}$ (call it Y), when we get $dY/\sqrt{Y^2+4}$ because $dX = YdY$. Conversely, at the places lying over infinity, the function does not have a pole but the differential does, since X is not a local parameter, but $1/X$ (call it Z) is, when the differential becomes $-dZ/Z*\sqrt{1-4Z^2}$, which has residues of ± 1 at the two places.

These residues form a 1-dimensional Z-module, so we wish to find a function with zeros of order ± 1 at the two places lying over infinity. Since Coates' Algorithm will not solve this problem directly, we divide the whole problem by X, and look for functions with zeros of order -1 at the two places lying over 0, and a zero of order 2 at one of the places lying over infinity (and a zero of order 0 at the other).

The initial basis is $\{1, \sqrt{X^2-4}\}$. We then have to divide each basis element by X in order to obtain an integral basis, and this is in fact a normal integral basis. Then a basis for the set of functions with poles no worse than those specified (i.e. - 1 at both the places lying over zero) is $\{1, 1/X, \sqrt{X^2-4}/X\}$. We want that linear combination of these functions which has a zero of order 2 at one of the places lying over infinity. This is , in fact, $(\sqrt{X^2-4} + X)/X$. We then have to multiply this by X (because we divided by X earlier) and we get $\sqrt{X^2-4} + X$ as the function with the correct zeros and poles. This computation took 403 milliseconds, 80 of which were spent doing the multiplying and dividing by X in order to transform the problem into one for which Coates' Algorithm was suitable.

We therefore have a logarithmic part of $\log (X + \sqrt{X^2-4})$, and in fact the derivative of this is the original integrand, so the problem is solved. The total time for this integration was 868 milliseconds, including all the time spent in Coates ' Algorithm. Repeated timing of this run at similar times of day[*] showed differences of about $\pm 3.5\%$ in the cpu time taken. The time taken when the number 4 was replaced by other perfect squares was very much the same, but when 4 was replaced by 2 the time increased by about 12%. This can be attributed to the fact that the program has to consider the possibility of there being residues at $\pm\sqrt{2}$ so it has to construct and then manipulate the algebraic expression $\sqrt{2}$. Supporting this view is the fact that the Coates' steps ran in almost exactly the same time.

If we now consider $1/\sqrt{2X^2-1}$, we have a slightly different computing behaviour, even though the integral is essentially of the same form. Here the residues actually are $\pm\sqrt{2}/2$, and all the Puiseux expansions involve $\sqrt{2}$ as well. This additional complexity means that the Coates step now takes 530 milliseconds, and the whole integration took 1590 milliseconds, or almost twice as long.

We can now present the output from the program which implements the work of the monograph, using a version which prints out a variety of intermediate results and observa-

---

[*] And therefore, it is hoped, a similar pattern of machine load. This is relevant to the timings quoted, because these are derived from operating system statistics, and the operating systems methods of accounting for time tend to allocate some of CPU time spent handling an interrupt to the task executing when the interrupt occurred. Therefore the timings quoted depend, to some extent, on the load on the machine.

tions.

```
INT(1/SQRT(X**2-4),X);
WITH 'NEW' FUNCTIONS :
((SQRT (PLUS (EXPT X 2) (MINUS 4))) X)
PLACES AT WHICH POLES COULD OCCUR
((QUOTIENT 1 X) (PLUS X 2) (PLUS X (MINUS 2)))
DIFFERENTIAL AFTER FIRST SUBSTITUTION IS


                   2
( - 1)/(SQRT( - 4*X  + 1)*X)

RESIDUES AT ((X QUOTIENT 1 X))
 ARE
((((X QUOTIENT 1 X)
     ((SQRT (PLUS (MINUS (TIMES 4 (EXPT X 2))) 1))
        SQRT
         (PLUS (MINUS (TIMES 4 (EXPT X 2))) 1)))
    -1
    . 1)
   (((X QUOTIENT 1 X)
       ((SQRT (PLUS (MINUS (TIMES 4 (EXPT X 2))) 1))
          MINUS
           (SQRT (PLUS (MINUS (TIMES 4 (EXPT X 2))) 1))))
        1
      . 1))
DIFFERENTIAL AFTER FIRST SUBSTITUTION IS

1/(SQRT(X - 4)*SQRT(X))

DIFFERENTIAL AFTER FIRST SUBSTITUTION IS

1/(SQRT(X + 4)*SQRT(X))

FIND FUNCTION WITH ZEROS OF ORDER:(-1 1)
 AT
(((X QUOTIENT 1 X)
    ((SQRT (PLUS (MINUS (TIMES 4 (EXPT X 2))) 1))
        SQRT
         (PLUS (MINUS (TIMES 4 (EXPT X 2))) 1)))
   ((X QUOTIENT 1 X)
      ((SQRT (PLUS (MINUS (TIMES 4 (EXPT X 2))) 1))
          MINUS
           (SQRT (PLUS (MINUS (TIMES 4 (EXPT X 2))) 1)))))
FIND FUNCTION WITH ZEROS OF ORDER:(-1 -1 2)
 AT
(((X . X)
```

```
   ((SQRT (PLUS (EXPT X 2) (MINUS 4)))
       SQRT
       (PLUS (EXPT X 2) (MINUS 4))))
  ((X . X)
     ((SQRT (PLUS (EXPT X 2) (MINUS 4)))
        MINUS
        (SQRT (PLUS (EXPT X 2) (MINUS 4)))))
  ((X QUOTIENT 1 X)
     ((SQRT (PLUS (MINUS (TIMES 4 (EXPT X 2))) 1))
        MINUS
        (SQRT (PLUS (MINUS (TIMES 4 (EXPT X 2))) 1)))))
SQRTS ON THIS CURVE :
((SQRT (PLUS (EXPT X 2) (MINUS 4)))
   (SQRT (PLUS (MINUS (TIMES 4 (EXPT X 2))) 1)))
INITIAL BASIS FOR THE SPACE M(X)

       2
SQRT(X  - 4)


1


INTEGRAL BASIS REDUCTION AT
((X . X))
MATRIX BEFORE A REDUCTION STEP:
%<(((((SQRT -1) . 1) . 2)) . 1),(((((SQRT -1) . 1) . -2)) . 1)%>
%<(1 . 1),(1 . 1)%>
NORMAL INTEGRAL BASIS REDUCTION WITH THE
 FOLLOWING SQRTS LYING OVER INFINITY :
((SQRT (PLUS (MINUS (TIMES 4 (EXPT X 2))) 1)))
EVALUATING
           2
SQRT( - 4*X  + 1)

%<0,0%>
%<(1 . 1),(-1 . 1)%>
EVALUATING
X

%<1,1%>
%<(1 . 1),(1 . 1)%>
MATRIX BEFORE A REDUCTION STEP AT INFINITY IS:
%<(1 . 1),(1 . 1)%>
%<(1 . 1),(-1 . 1)%>
BASIS FOR THE FUNCTIONS WITH PRECISELY THE CORRECT POLES

       2
SQRT(X  - 4)/X
```

1

1/X

EQUATIONS TO BE SOLVED:
ROW NUMBER 0
 - 1

1

0

ROW NUMBER 1
0

0

1

ROW NUMBER 2
1

1

1

ANSWER FROM LINEAR EQUATION SOLVING IS

$$(SQRT(X^2 - 4) + X)/X$$

(COATES TIME 396 MILLISECONDS)
EXTENSION LOGARITHM IS

$$SQRT(X^2 - 4) + X$$

INNER WORKING YIELDS

$$LOG(SQRT(X^2 - 4) + X)$$

WITH DERIVATIVE

$$(SQRT(X^2 - 4)*X + X^2 - 4)/(lastsqrt*X^2 - 4*lastsqrt + X^3 - 4*X)$$

```
(TIME TAKEN 868 MILLISECONDS)
```

$$LOG(SQRT(X^2 - 4) + X)$$

Example 2: 1/SQRT((X**2-1)*(X**2-K**2)).

This is the typical elliptic integral, which is in fact known to be unintegrable in elementary terms unless $K=1$. The only places where $dX/\sqrt{(X^2-1)(X^2-K^2)}$ could have poles are infinity and the places lying over ± 1 and $\pm K$. Considering first the place at infinity, let us write $Y = 1/X$, so that Y is a local parameter at infinity. This integrand is then $(-dY/Y^2)/\sqrt{(Y^{-2}-1)(Y^{-2}-K^2)} = -dY/\sqrt{(1-Y^2)(1-K^2Y^2)}$, and this clearly is finite at $Y=0$ (i.e. at X infinite). At one of the other places (i.e. the roots of the radical), let Y be X-root. Then the integrand is $dY/(\sqrt{Y}(\text{product of other 3 terms})$. The product of the other terms has neither a pole nor a zero at $Y=0$ because all the roots are distinct (which explains why $K=\pm 1$ is a special case). Now let $Z^2 = Y$, so that Z is a local parameter (Y was not since we had $\sqrt{Y}$ in our expression). The integrand is then $(2ZdZ)/(Z\sqrt{\text{product}}) = 2dZ/\sqrt{\text{product}}$, which does not have a pole at $Z=0$.

Since the integrand has no poles, it can have no residues, and therefore the integral (if it is elementary) has no logarithmic part, i.e. it must be purely algebraic. This algebraic function can have no finite poles (else its derivative would have, and we have just shown that it does not) and must have a pole of order at most 1 at infinity. Hence the integral (if it is elementary) is of the form $a+bX$, and on differentiating this it is easy to see that this is not correct. Therefore this function does not have an elementary integral.

Example 3: A Torsion Example.

Consider the integrand $3YX^2/2(Y + 1)(X^3 + 1)$ with $Y = \sqrt{X^3 + 1}$. This integrand is chosen to be the derivative of $LOG(1+Y)$, to demonstrate how apparently trifling integrals can require sophisticated machinery.

The integrand could have residues at infinity, 0 or any of the roots of

$Y^2 = X^3 + 1 = 0$. In fact the integrand has a residue of -3 at infinity, which is a ramified point with ramification index 2, and a residue of 3 at one place lying over 0, but not at the other. On the locally non-singular model in which infinity is not ramified, there are two points lying over infinity, at each of which the integrand has residue $-3/2$. Hence we consider r_1 to be $3/2$, and have the divisor D consisting of one of the places lying over 0 (hereafter termed the place $0+$) with multiplicity 2, and the one ramified place at infinity with multiplicity -1. In terms of a non-singular model (the curve $Y^2 = X^6 + 1$) we have $0+$ with multiplicity 2 at two places lying over infinity (say $I+$ and $I-$) with multiplicity -1 each.

The divisor D is in fact not linearly equivalent to 0 by Coates Algorithm, since the only function with poles of order -1 at $I+$ and $I-$ but nowhere else is X, and this clearly cannot be made to have a zero of order 2 at $0+$. Since we are looking for functions with poles at infinity, we are in fact not able to use Coates' Algorithm directly, but have to consider the problem of finding functions with a pole of order 1 at $0-$ and a zero of order 1 at $0+$, and then multiply these functions by X in order to regain functions with the poles and zeros of the divisor D.

Since $Y^2 = X^3 + 1$ is an elliptic curve (it has one linearly independent differential of the first kind, viz $1/Y$) defined over the rationals, and since D is defined over the rationals, we can use Mazur's[$] bound on the torsion of elliptic curves over the rationals (see Chapter 5). Hence we have to consider $2D, 3D, ..., 10D, 12D$ in turn until we find one which is the divisor of a function. $2D$ is not, but $3D$ is in fact the divisor of the function $2Y + X^3 + 2$.

Hence the logarithmic part is in fact

[$] Note that the similar problem, with $Y^2 = X^3 + 8$, which can easily be transformed into the other, would not be amenable to Mazur's bound since then the divisor would cease to be defined over the rationals, even though the curve would still be defined over the rationals. We would need to extend our constant field by the square root of 2. Such a problem could be solved by means of the LUTZ__NAGELL Algorithm (see Chapter 7) or we could attempt to bound the torsion for this elliptic curve over the extended field. This particular case is discussed as an example of operating over algebraic number fields in Chapter 8.

$$\frac{3}{2}\frac{1}{3} \log (2Y + X^3 + 2) = \frac{\log (2Y + X^3 + 2)}{2}.$$

When we differentiate this, we find that it is equal to the original integrand, and hence there is no algebraic part. Note that we have not ended up with quite the expression we originally differentiated, but the two expressions in fact represent the same function. This computation took 5191 milliseconds on the Cambridge 370/165 (and 3.97 seconds in garbage collection time), of which the three Coates' Algorithm steps consumed 400, 403, and 530 milliseconds respectively.

Example 4: Modular Curve $X(7)$

As remarked by Tate (1974, p.198), the point $(0,0)$ is of order 7 on the curve $y^2 + (1 + d - d^2)xy + (d^2 - d^3)y = x^3 + (d^2 - d^3)x^2$ which is elliptic if its discriminant $d^7(d-1)^7(d^3 - 8d^2 + 5d + 1)$ is non-zero. We can convert this into a more standard form, when the assertion becomes that the point $P = (0, d^3 - d^2)$ is of order 7 on the curve $y^2 = x^3 + x^2(1 + 2d + 3d^2 - 6d^3 + d^4) + 4xd^2(1 - 4d^2 + 2d^3) + 16d^4(1-d)^2$ which is almost a Weierstrass canonical form. The point $-2P$ is $(d(d-1), d(d-1)^3)$, and this is also of order 7 by elementary group theory. The divisor consisting of this point with order -7, and infinity with order 7, is therefore principal, and in fact corresponds to the function

$$G(x,y) = x^{-7}$$

$$\{yd^2 + 2ydx + yx^2 + yx + d^5 + d^4(3x-1) + d^3(3x^2 - 2x)$$

$$+ d^2(x^3 - 4x^2 - 2x) + d(-3x^3 - 3x^2) - 3x^3 - x^2)\}$$

as can be seen by applying Coates' Algorithm[*] to this divisor.

We therefore wish to consider integrating the derivative of log G, which is (as computed by REDUCE-2[#]):

```
           6      5        5    4 2      4      4     3 2
( - 7*SQRT(D  + 2*D *X - 2*D  + D *X  - 4*D *X + D  - 6*D *X


     2 2      2         2       3    2    8
 + 3*D *X  + 2*D *X + 2*D*X  + 4*X  + X )*D  - 25*lastsqrt*


 7                7              6 2                  6
D *X + 14*lastsqrt*D  - 33*lastsqrt*D *X  + 44*lastsqrt*D *X


               6              5 3                5 2
 - 7*lastsqrt*D  - 19*lastsqrt*D *X  + 79*lastsqrt*D *X  - 4


         4 4              4 3              4 2
*lastsqrt*D *X  + 73*lastsqrt*D *X  - lastsqrt*D *X  - 19*


         4              3 4              3 2
lastsqrt*D *X + 24*lastsqrt*D *X  - 34*lastsqrt*D *X  - 12*


         2 4              2 3              2 2
lastsqrt*D *X  - 66*lastsqrt*D *X  - 17*lastsqrt*D *X  - 44*
```

---

[*] This provides another example of the great computational difficulties that surround this subject. The application of Coates' Algorithm to find the functions took 417 seconds (+ 184 seconds garbage collecting) in 650K on the Cambridge 370/165. Solving for the zeros as well (i.e. a 14 by 7 set of linear equations) was not complete in a further 517 seconds, and furthermore filled all the available store with numbers of the order of 10^{2400} as coefficients of intermediate expressions. A rewritten version of the linear equation solver which looked for sparse rows first and treated them specially was able to solve the linear equations in essentially 0 time. This is related to the points made by Wang and Minamikawa (1976) when they compared Bareiss-type fraction-free methods (Lipson,1969) similar to the previous implementation, with ones which looked for such structure.

[#] The version of REDUCE-2 on which the integration system described here is based has a local modification (see Appendix 1), which causes the symbol "lastsqrt" to be printed in place of a SQRT which is identical to the immediately preceding SQRT in the same expression. The reader can imagine what the expression below would look like if this device were not adopted.

$$\text{lastsqrt*D*X}^4 - 20\text{*lastsqrt*D*X}^3 - 14\text{*lastsqrt*X}^5 - 22*$$

$$\text{lastsqrt*X}^4 - 5\text{*lastsqrt*X}^3 - 7\text{*D}^{11} - 32\text{*D}^{10}\text{*X} + 21\text{*D}^{10} - 58$$

$$\text{*D}^9\text{*X}^2 + 90\text{*D}^9\text{*X} - 21\text{*D}^9 - 52\text{*D}^8\text{*X}^3 + 195\text{*D}^8\text{*X}^2 - 58\text{*D}^8\text{*X} +$$

$$7\text{*D}^8 - 23\text{*D}^7\text{*X}^4 + 240\text{*D}^7\text{*X}^3 - 127\text{*D}^7\text{*X}^2 - 26\text{*D}^7\text{*X} - 4\text{*D}^6\text{*X}^5$$

$$+ 150\text{*D}^6\text{*X}^4 - 180\text{*D}^6\text{*X}^3 - 82\text{*D}^6\text{*X}^2 + 26\text{*D}^6\text{*X} + 36\text{*D}^5\text{*X}^5 -$$

$$174\text{*D}^5\text{*X}^4 - 192\text{*D}^5\text{*X}^3 + 36\text{*D}^5\text{*X}^2 - 72\text{*D}^4\text{*X}^5 - 183\text{*D}^4\text{*X}^4 +$$

$$102\text{*D}^4\text{*X}^3 + 36\text{*D}^4\text{*X}^2 - 104\text{*D}^3\text{*X}^5 + 112\text{*D}^3\text{*X}^4 + 66\text{*D}^3\text{*X}^3 - 16$$

$$\text{*D}^2\text{*X}^6 + 136\text{*D}^2\text{*X}^5 + 137\text{*D}^2\text{*X}^4 + 22\text{*D}^2\text{*X}^3 + 48\text{*D*X}^6 + 96\text{*D*}$$

$$\text{X}^5 + 25\text{*D*X}^4 + 48\text{*X}^6 + 32\text{*X}^5 + 5\text{*X}^4)/(X\text{*(lastsqrt*D}^8 + 4*$$

$$\text{lastsqrt*D}^7\text{*X} - 2\text{*lastsqrt*D}^7 + 6\text{*lastsqrt*D}^6\text{*X}^2 - 7*$$

$$\text{lastsqrt*D}^6\text{*X} + \text{lastsqrt*D}^6 + 4\text{*lastsqrt*D}^5\text{*X}^3 - 14*$$

$$\text{lastsqrt*D}^5\text{*X}^2 + \text{lastsqrt*D}^4\text{*X}^4 - 15\text{*lastsqrt*D}^4\text{*X}^3 + 3$$

$$\text{*lastsqrt*D}^4\text{*X} - 6\text{*lastsqrt*D}^3\text{*X}^4 + 6\text{*lastsqrt*D}^3\text{*X}^2 +$$

$$3\text{*lastsqrt*D}^2\text{*X}^4 + 13\text{*lastsqrt*D}^2\text{*X}^3 + 3\text{*lastsqrt*D}^2\text{*X}^2$$

$$+ 10\text{*lastsqrt*D*X}^4 + 4\text{*lastsqrt*D*X}^3 + 4\text{*lastsqrt*X}^5$$

$$+ 5\text{*lastsqrt*X}^4 + \text{lastsqrt*X}^3 + \text{D}^{11} + 5\text{*D}^{10}\text{*X} - 3\text{*D}^{10}$$

```
        9  2       9          9         8  3        8  2
+ 10*D *X   - 14*D *X + 3*D   + 10*D *X   - 33*D *X   + 9*

  8       8       7  4        7  3        7  2       7       6
D *X -  D  + 5*D *X   - 45*D *X   + 21*D *X   + 4*D *X + D

  5       6  4        6  3        6  2       6        5  5
*X   - 32*D *X   + 33*D *X   + 14*D *X   - 4*D *X - 9*D *X

        5  4        5  3        5  2       4  5        4  4
+ 36*D *X   + 35*D *X   - 6*D *X   + 18*D *X   + 37*D *X

        4  3       4  2        3  5        3  4        3  3
- 18*D *X   - 6*D *X   + 23*D *X   - 22*D *X   - 12*D *X

        2  6        2  5        2  4       2  3        6
+ 4*D *X   - 30*D *X   - 27*D *X   - 4*D *X   - 12*D*X   -

     5        4        6        5       4
21*D*X   - 5*D*X   - 12*X   - 7*X   - X ))
```

This enormous integrand unfortunately stretches the machine resources available too far; computing the residues took[#] 34 seconds, and the initial Coates' step (i.e. seeing if the divisor is principal) took 6 seconds, the computation of the differentials of the first kind took 2 seconds,[*] the second Coates' step (i.e. working on 2D) took 18 seconds, and the third Coates' step ran out of time after 200 seconds.

In the particular case of $d=2$, the above assertions reduce to the point $(0,4)$ being of order 7 on the curve $y^2 = 4x^3 - 15x^2 + 8x + 16$. Hence the divisor consisting on this point P with order -1, and the point at infinity O with order 1, is a divisor of order 7. We can use Coates' Algorithm to find a function which corresponds to 7 times this divisor, and in fact the function $G(x,y) = (-yx^2 - 5yx - 4y + 5x^3 - x^2 - 24x - 16)/x^7$ has the property that it has a pole of order 7 at P, and a zero of order 7 at the point at infinity (there is one ramified point at infinity).

[#] All these timings were taken from jobs run on on the Cambridge 370/165 in 700k. Garbage collection time is not included, but averaged an extra 20%.

[*] The FIND_ORDER_MANIN work was not done in this run, since it had been computed earlier (see chapter 6) and would have made the whole job prohibitively expensive.

We can then differentiate this function and try re-integrating it to test the accuracy of our algebraic geometry, at least on curves of genus 1. We note that the curve is elliptic over the rationals, so that we can apply Mazur's bound on the torsion. In fact we can also apply Cayley's determinant test, and, as described in Chapter 5, this leads to a substantial saving in the total computational cost of this integral, which is still substantial at 40 seconds CPU (+ 20 seconds spent garbage collecting when running in 550K bytes on the IBM 370/165).

Example 5: Chebyshev's Integral

In his paper (Chebyshev, 1857) on continued fractions and the integration of pseudo-elliptic functions, Chebyshev considered the following integral:

$$\int \frac{2X^6+4X^5+7X^4-3X^3-X^2-8X-8}{(2X^2-1)^2\sqrt{X^4+4X^3+2X^2+1}}\,dX.$$

This integrand has 4 residues, of magnitude $5/2$ (but naively computed as SQRT(4*SQRT(2)+9)*(10*SQRT(2)-5)/14, unless one notices that 4*SQRT(2)+9 is a perfect square in $Q(\sqrt{2})$ - see Chapter 2 for details of these problems). The divisor in fact turns out to be of order 5, and $5D$ is the divisor of

$$\frac{A(X)Y-B(X)}{C(X)},$$

where $A(X) = 1023X^8 + 4104X^7 + 5084X^6 + 2182X^5 + 805X^4 + 624X^3 + 10X^2 + 28X$, and $B(X) = 1025X^{10} + 6138X^9 + 12307X^8 + 10188X^7 + 4503X^6 + 3134X^5$
$$+ 1589X^4 + 140X^3 + 176X^2 + 2,$$
$C(X) = 32X^{10}-80X^8 + 80X^6-40X^4 + 10X^2-1$, and $Y = \sqrt{X^4+4X^3+2X^2+1}$.

Although the curve defined by X and Y is elliptic, and the defining equation has rational coefficients, we cannot apply Mazur's bound, because the residues of the integrand, and hence the divisor in question, lies over $X = 2^{-1/2}$ and $X = -2^{-1/2}$, and hence the divisor is not defined over the rationals. At a time when the theory of Chapter 7 had not been implemented, there was no program available to bound the torsion, so we had to settle for a partial computation, i.e. one that will terminate with the integral if the function is integrable, but may never terminate if it is not.

We therefore tried every possible order for this divisor in turn, and compute that D, $2D$, $3D$, $4D$ are not linearly equivalent to 0, but that $5D$ is. The whole process (which did not use Cayley's technique as described in chapter 5), took 75.6 seconds (+ 35.8 seconds garbage collection time) on the Cambridge 370/165 in 650Kbytes. It is interesting to note that 30 seconds of this was spent in solving the linear equations (see also the footnote to the previous example) which define that linear combination of the elements of our normal integral basis which has all the correct zeros.

This gives us a solution of

$$\frac{(2X+1)Y}{2(2X^2-1)} + \frac{\log\left(\dfrac{A(X)Y-B(X)}{C(X)}\right)}{2},$$

where A, B and C are as defined on the previous page, and $Y^2 = X^4 + 4X^3 + sX^2 + 1$. Note that Chebyshev's own solution has the same algebraic part, but his logarithmic part has coefficient $1/4$, and hence the argument of his logarithm is essentially the square of ours.

When the LUTZ__NAGELL code of Chapter 7 was implemented, it readily found that the point which was the sum of the terms in the divisor was of order 5 (because $4P = -P$) and hence we needed merely to try $5D$. It is interesting to note that this computation in fact took significantly longer than that when the LUTZ__NAGELL code was not available, though it is not clear to what extent this merely reflects a difference in the amount of tuning that has gone into the two pieces of code. Of course, the LUTZ__NAGELL code has the great advantage of certainty, in that had the divisor been of infinite order, we would have been told so.

This example was re-run on the REDUCE version of 15th. April 1979, which supports the printing of sub-expressions, and we attach the output from this run to demonstrate the operation of the program on a complicated case. This was run[*] with a lower level of tracing than the output in Example 1 above, since otherwise the results would be even lengthier than they now are. We first give the output from running the program with DIVISOR__TO__FUNCTION implemented as a call of Coates with all the poles, followed by implementing all the zeros as linear constraints.

```
INT((2*X**6+4*X**5+7*X**4-3*X**3-X**2-8*X-8)*(2*X**2-1)**(-2)/
    SQRT(X**4+4*X**3+2*X**2+1),X);
PLACES AT WHICH POLES COULD OCCUR
X=infinity
X=-1
X=Any root of:
     3       2
   - X   - 3*X   + 2*X - 1

X=
( - SQRT(2))/2

X=
SQRT(2)/2

NEW SET OF RESIDUES ARE
( - 5)/2

5/2

( - 5)/2

5/2
FIND FUNCTION WITH ZEROS OF ORDER:(1 -1 1 -1)
PROVED INSOLUBLE MOD 103
(COATES TIME 4020 MILLISECONDS)
FIND THE DIFFERENTIALS OF THE FIRST KIND ON CURVE DEFINED BY:
   4       3       2
X   + 4*X   + 2*X   + 1
DIFFERENTIALS ARE:
1/ANS1
    WHERE
```

---

[*] In 700K bytes, and took 105 seconds with an extra 75 seconds spent on garbage collection and program loading.

$$\text{ANS1} := \text{SQRT}(X^4 + 4*X^3 + 2*X^2 + 1)$$
FIND WEIERSTRASS FORM FOR ELLIPTIC CURVE DEFINED BY:
ANS1
   WHERE
$$\text{ANS1} := \text{SQRT}(X^4 + 4*X^3 + 2*X^2 + 1)$$
FIND FUNCTION WITH ZEROS OF ORDER: (-3)
FUNCTION IS
$$(2*\text{ANS1}*X + \text{ANS1} + 2*X^3 + 5*X^3)/X$$
   WHERE
$$\text{ANS1} := \text{SQRT}(X^4 + 4*X^3 + 2*X^2 + 1)$$
FIND FUNCTION WITH ZEROS OF ORDER: (-2)
FUNCTION IS
$$(\text{ANS1} + X^2 + 2*X)/X^2$$
   WHERE
$$\text{ANS1} := \text{SQRT}(X^4 + 4*X^3 + 2*X^2 + 1)$$
STANDARD FORM IS Y**2 =
$$X^3 - 12*X + 38$$
POINT TO HAVE TORSION INVESTIGATED IS
(-2)

$$- 3*\text{SQRT}(2)*\text{SQRT}(3)$$
POINT IS OF ORDER DIVIDING 5
POINT ACTUALLY IS OF ORDER 5
FIND FUNCTION WITH ZEROS OF ORDER: (5 -5 5 -5)
(COATES TIME 72940 MILLISECONDS)
DIVISOR HAS ORDER 5
EXTENSION LOGARITHM IS
ANS1
   WHERE
$$\text{ANS1} := \text{NTHROOT}((1023*\text{SQRT}(X^4 + 4*X^3 + 2*X^2 + 1)*X^8 + 4104*$$
$$\text{SQRT}(X^4 + 4*X^3 + 2*X^2 + 1)*X^7 + 5084*\text{SQRT}(X^4 + 4*X^3 + 2*X^2 + 1)*$$
$$X^6 + 2182*\text{SQRT}(X^4 + 4*X^3 + 2*X^2 + 1)*X^5 + 805*\text{SQRT}(X^4 + 4*X^3 + 2$$
$$*X^2 + 1)*X^4 + 624*\text{SQRT}(X^4 + 4*X^3 + 2*X^2 + 1)*X^3 + 10*\text{SQRT}(X^4 + 4$$
$$*X^3 + 2*X^2 + 1)*X^2 + 28*\text{SQRT}(X^4 + 4*X^3 + 2*X^2 + 1)*X - 1025*X^{10}$$

$$- 6138X^9 - 12307X^8 - 10188X^7 - 4503X^6 - 3134X^5 - 1589X^4$$
$$- 140X^3 - 176X^2 - 2)/(32X^{10} - 80X^8 + 80X^6 - 40X^4 + 10X^2$$
$$- 1),5)$$

INNER WORKING YIELDS

$$(- ANS1 + ANS2)/2$$

  WHERE

$$ANS2 := LOG(1023\sqrt{X^4 + 4X^3 + 2X^2 + 1}\,X^8 + 4104\sqrt{X^4 + 4X^3 + 2X^2 + 1}\,X^7 + 5084\sqrt{X^4 + 4X^3 + 2X^2 + 1}\,X^6 + 2182\sqrt{X^4 + 4X^3 + 2X^2 + 1}\,X^5 + 805\sqrt{X^4 + 4X^3 + 2X^2 + 1}\,X^4 + 624\sqrt{X^4 + 4X^3 + 2X^2 + 1}\,X^3 + 10\sqrt{X^4 + 4X^3 + 2X^2 + 1}\,X^2 + 28\sqrt{X^4 + 4X^3 + 2X^2 + 1}\,X - 1025X^{10} - 6138X^9 - 12307X^8 - 10188X^7 - 4503X^6 - 3134X^5 - 1589X^4 - 140X^3 - 176X^2 - 2)$$

$$ANS1 := LOG(32X^{10} - 80X^8 + 80X^6 - 40X^4 + 10X^2 - 1)$$

WITH DERIVATIVE

$$(- 6150\,ANS1\,X^{12} - 41953\,ANS1\,X^{11} - 111707\,ANS1\,X^{10} - 165629\,ANS1\,X^9 - 164107\,ANS1\,X^8 - 112635\,ANS1\,X^7 - 56725\,ANS1\,X^6 - 30723\,ANS1\,X^5 - 12879\,ANS1\,X^4 - 1860\,ANS1\,X^3 - 1244\,ANS1\,X^2 - 10\,ANS1\,X - 14\,ANS1 + 6138X^{14} + 54291X^{13} + 189417X^{12} + 359458X^{11} + 442796X^{10} + 392574X^9 + 261828X^8 + 153172X^7 + 87512X^6 + 33251X^5 + 10019X^4 + 4978X^3 + 210X^2 + 196X)/(2046\,ANS1\,X^{14} + 16392\,ANS1\,X^{13} + 46069\,ANS1\,X^{12} + 53256\,ANS1\,X^{11} + 17902\,ANS1\,X^{10} - 6102\,ANS1\,X^9 - 2324\,ANS1\,X^8 - 5316\,ANS1\,X^7 - 7326\,ANS1\,X^6 - 2138\,ANS1\,X^5 - 917\,ANS1\,X^4 - 624\,ANS1\,X^3 - 10\,ANS1\,X^2 - 28\,ANS1\,X - 2050X^{16} - 20476X^{15} - 76793X^{14} - 133146X^{13} - 102879X^{12} - 23628X^{11} + 18X^{10} + 1756X^9 + 18604X^8 + 14716X^7 + 4531X^6 + 3822X^5 + 1583X^4 + 148X^3 + 176X^2 + 2)$$

WHERE

$$\text{ANS1} := \text{SQRT}(X^4 + 4*X^3 + 2*X^2 + 1)$$

POTENTIAL CANCELLATION DETECTED
(TIME TAKEN 108473 MILLISECONDS)

$$(- 2*\text{LOG}(32*X^{10} - 80*X^8 + 80*X^6 - 40*X^4 + 10*X^2 - 1)*X^2 + \text{LOG}(32*$$
$$X^{10} - 80*X^8 + 80*X^6 - 40*X^4 + 10*X^2 - 1) + 2*\text{LOG}(1023*\text{SQRT}(X^4 + 4*$$
$$X^3 + 2*X^2 + 1)*X^8 + 4104*\text{SQRT}(X^4 + 4*X^3 + 2*X^2 + 1)*X^7 + 5084*\text{SQRT}$$
$$(X^4 + 4*X^3 + 2*X^2 + 1)*X^6 + 2182*\text{SQRT}(X^4 + 4*X^3 + 2*X^2 + 1)*X^5 +$$
$$805*\text{SQRT}(X^4 + 4*X^3 + 2*X^2 + 1)*X^4 + 624*\text{SQRT}(X^4 + 4*X^3 + 2*X^2 + 1)$$
$$*X^3 + 10*\text{SQRT}(X^4 + 4*X^3 + 2*X^2 + 1)*X^2 + 28*\text{SQRT}(X^4 + 4*X^3 + 2*X^2$$
$$+ 1)*X - 1025*X^{10} - 6138*X^9 - 12307*X^8 - 10188*X^7 - 4503*X^6 -$$
$$3134*X^5 - 1589*X^4 - 140*X^3 - 176*X^2 - 2)*X^2 - \text{LOG}(1023*\text{SQRT}(X^4 + 4$$
$$*X^3 + 2*X^2 + 1)*X^8 + 4104*\text{SQRT}(X^4 + 4*X^3 + 2*X^2 + 1)*X^7 + 5084*$$
$$\text{SQRT}(X^4 + 4*X^3 + 2*X^2 + 1)*X^6 + 2182*\text{SQRT}(X^4 + 4*X^3 + 2*X^2 + 1)*X^5$$
$$+ 805*\text{SQRT}(X^4 + 4*X^3 + 2*X^2 + 1)*X^4 + 624*\text{SQRT}(X^4 + 4*X^3 + 2*X^2$$
$$+ 1)*X^3 + 10*\text{SQRT}(X^4 + 4*X^3 + 2*X^2 + 1)*X^2 + 28*\text{SQRT}(X^4 + 4*X^3 +$$
$$2*X^2 + 1)*X - 1025*X^{10} - 6138*X^9 - 12307*X^8 - 10188*X^7 - 4503*X^6$$
$$- 3134*X^5 - 1589*X^4 - 140*X^3 - 176*X^2 - 2) + 2*\text{SQRT}(X^4 + 4*X^3 + 2$$
$$*X^2 + 1)*X + \text{SQRT}(X^4 + 4*X^3 + 2*X^2 + 1))/(2*(2*X^2 - 1))$$

When this example is run using the algorithm **DIVISOR__TO__FUNCTION** as described in chapter 3, the answers are very different. The output from this run is given below, and it is noticeable that, while the time for the operation of Coates' algorithm on the divisor $5D$ is much decreased (to about 37.3% of its former value), the total time is increased by about 12.3%, which is just accountable for by random fluctuations. This is mainly caused by the more complicated form of the logarithmic part, which means that differentiating it took 10 seconds, and subtracting it from the original integral (which includes placing the two over a common denominator) took 20. The form of the logarithmic part is very interesting, though it is not clear whether or not this form is better than the one enormous logarithm of the previous run.

```
Y:=(2*X**6+4*X**5+7*X**4-3*X**3-X*X-8*X-8)/
    ((2*X**2-1)**2*SQRT(X**4+4*X**3+2*X**2+1));
          6     5     4     3    2                    4      3
Y := (2*X  + 4*X  + 7*X  - 3*X  - X  - 8*X - 8)/(SQRT(X  + 4*X  + 2
        2          4     2
      *X  + 1)*(4*X  - 4*X  + 1))
INT(Y,X);
PLACES AT WHICH POLES COULD OCCUR
X=infinity
X=-1
X=Any root of:
     3      2
 - X  - 3*X  + 2*X - 1

X=
( - SQRT(2))/2

X=
SQRT(2)/2

NEW SET OF RESIDUES ARE
( - 5)/2

5/2

( - 5)/2

5/2
FIND FUNCTION WITH ZEROS OF ORDER:(1 1 -1 -1)
PROVED INSOLUBLE MOD 103
(COATES TIME 4403 MILLISECONDS)
FIND THE DIFFERENTIALS OF THE FIRST KIND ON CURVE DEFINED BY:
 4     3     2
X  + 4*X  + 2*X  + 1
DIFFERENTIALS ARE:
1/ANS1
   WHERE
                     4     3     2
     ANS1 := SQRT(X  + 4*X  + 2*X  + 1)
FIND WEIERSTRASS FORM FOR ELLIPTIC CURVE DEFINED BY:
```

```
ANS1
   WHERE
                    4        3       2
      ANS1 := SQRT(X  + 4*X  + 2*X  + 1)
FIND FUNCTION WITH ZEROS OF ORDER:(-3)
FUNCTION IS
                         3      2    3
(2*ANS1*X + ANS1 + 2*X  + 5*X )/X
   WHERE
                    4        3       2
      ANS1 := SQRT(X  + 4*X  + 2*X  + 1)
FIND FUNCTION WITH ZEROS OF ORDER:(-2)
FUNCTION IS
         2          2
(ANS1 + X  + 2*X)/X
   WHERE
                    4        3       2
      ANS1 := SQRT(X  + 4*X  + 2*X  + 1)
STANDARD FORM IS Y**2 =
 3
X  - 12*X + 38
POINT TO HAVE TORSION INVESTIGATED IS
(-2)

 - 3*SQRT(2)*SQRT(3)
POINT IS OF ORDER DIVIDING 5
POINT ACTUALLY IS OF ORDER 5
OPERATE ON DIVISOR:(5 5 -5 -5)
AT
X=
SQRT(2)/2
AT THE PLACE +
X=
( - SQRT(2))/2
AT THE PLACE +
X=
SQRT(2)/2
AT THE PLACE -
X=
( - SQRT(2))/2
AT THE PLACE -
FIND FUNCTION WITH ZEROS OF ORDER:(-1 -1 1)
REPLACED BY THE POLE
X=
 - SQRT(2) + 2
AT THE PLACE -
5 TIMES
```

FIND FUNCTION WITH ZEROS OF ORDER:(-2 1)
REPLACED BY THE POLE
X=
SQRT(2) + 2
AT THE PLACE +
2 TIMES
FIND FUNCTION WITH ZEROS OF ORDER:(-1 -1 1)
REPLACED BY THE POLE
X=
SQRT(2)/2
AT THE PLACE -
1 TIMES
FIND FUNCTION WITH ZEROS OF ORDER:(2 -1 -1)
(COATES TIME 27197 MILLISECONDS)
DIVISOR HAS ORDER 5
EXTENSION LOGARITHM IS
ANS1
   WHERE

$$
\begin{aligned}
\text{ANS1} := \text{NTHROOT}((&-731{*}\text{SQRT}(X^4 + 4{*}X^3 + 2{*}X^2 + 1){*}\text{SQRT}(2){*}X^4 \\
&-71492{*}\text{SQRT}(X^4 + 4{*}X^3 + 2{*}X^2 + 1){*}\text{SQRT}(2) + 70030{*}\text{SQRT}(X^4 + 4{*} \\
&X^3 + 2{*}X^2 + 1){*}X + 141522{*}\text{SQRT}(X^4 + 4{*}X^3 + 2{*}X^2 + 1) + 40957{*} \\
&\text{SQRT}(2){*}X^3 + 174250{*}\text{SQRT}(2){*}X^2 + 122871{*}\text{SQRT}(2){*}X - 50648{*}\text{SQRT}(2 \\
&) - 90874{*}X^3 - 403722{*}X^2 - 272622{*}X + 61070)/(\text{SQRT}(2){*}X^2 + 4{*} \\
&\text{SQRT}(2){*}X + 2{*}\text{SQRT}(2) + 2{*}X^3 + 8{*}X^2 + 4{*}X){*}(-2{*}\text{SQRT}(X^4 + 4{*}X^3 \\
&+ 2{*}X^2 + 1) + \text{SQRT}(2){*}X + 2{*}\text{SQRT}(2))/(X^2 + 4{*}X + 2){*} \\
&(10{*}\text{SQRT}(X^4 + 4{*}X^3 + 2{*}X^2 + 1){*}\text{SQRT}(2) + 17{*}\text{SQRT}(X^4 + 4{*}X^3 + 2{*}X^2 \\
&+ 1) + 4{*}\text{SQRT}(2){*}X^2 + 16{*}\text{SQRT}(2){*}X - 2{*}\text{SQRT}(2) - 11{*}X^2 - 44{*}X - \\
&39)/(2{*}\text{SQRT}(2){*}X + 4{*}\text{SQRT}(2) - X^2 - 4{*}X - 6){*} \\
&(-4{*}\text{SQRT}(X^4 + 4{*}X^3 + 2{*}X^2 + 1){*}\text{SQRT}(2) - 19{*}\text{SQRT}(X^4 + 4{*}X^3 + 2{*}X^2 \\
&+ 1) + 16{*}\text{SQRT}(2){*}X^2 + 8{*}\text{SQRT}(2){*}X - 6{*}\text{SQRT}(2) + 29{*}X^2 + 38{*}X - 5 \\
&)/(2{*}X^2 - 1)^5 , 5)
\end{aligned}
$$

INNER WORKING YIELDS
$(6{*}\text{LOG}((-1)) - 2{*}\text{ANS1} - \text{ANS2} + \text{ANS3} - 5{*}\text{ANS4} - 2{*}\text{ANS5} + 5{*}\text{ANS6} + 2{*}$
$\text{ANS7} + \text{ANS8})/2$

WHERE

ANS8 := LOG(731*SQRT(2)*SQRT(X^4 + 4*X^3 + 2*X^2 + 1)*X + 71492*SQRT(2)*SQRT(X^4 + 4*X^3 + 2*X^2 + 1) - 40957*SQRT(2)*X^3 - 174250*SQRT(2)*X^2 - 122871*SQRT(2)*X + 50648*SQRT(2) - 70030*SQRT(X^4 + 4*X^3 + 2*X^2 + 1)*X - 141522*SQRT(X^4 + 4*X^3 + 2*X^2 + 1) + 90874*X^3 + 403722*X^2 + 272622*X - 61070)

ANS7 := LOG(10*SQRT(2)*SQRT(X^4 + 4*X^3 + 2*X^2 + 1) + 4*SQRT(2)*X^2 + 16*SQRT(2)*X - 2*SQRT(2) + 17*SQRT(X^4 + 4*X^3 + 2*X^2 + 1) - 11*X^2 - 44*X - 39)

ANS6 := LOG(4*SQRT(2)*SQRT(X^4 + 4*X^3 + 2*X^2 + 1) - 16*SQRT(2)*X^2 - 8*SQRT(2)*X + 6*SQRT(2) + 19*SQRT(X^4 + 4*X^3 + 2*X^2 + 1) - 29*X^2 - 38*X + 5)

ANS5 := LOG(2*SQRT(2)*X + 4*SQRT(2) - X^2 - 4*X - 6)

ANS4 := LOG(2*X^2 - 1)

ANS3 := LOG(SQRT(2)*X + 2*SQRT(2) - 2*SQRT(X^4 + 4*X^3 + 2*X^2 + 1))

ANS2 := LOG(SQRT(2) + 2*X)

ANS1 := LOG(X^2 + 4*X + 2)

WITH DERIVATIVE

(ANS1*(6*X^2 + 5*X + 7))/(2*X^6 + 8*X^5 + 3*X^4 - 4*X^3 - 1)

  WHERE

ANS1 := SQRT(X^4 + 4*X^3 + 2*X^2 + 1)

POTENTIAL CANCELLATION DETECTED

(TIME TAKEN 121778 MILLISECONDS)

(- 4*LOG(X^2 + 4*X + 2)*X^2 + 2*LOG(X^2 + 4*X + 2) - 2*LOG(SQRT(2) + 2*X)*X^2 + LOG(SQRT(2) + 2*X) + 2*LOG(SQRT(2)*X + 2*SQRT(2) - 2*SQRT(X^4 + 4*X^3 + 2*X^2 + 1))*X^2 - LOG(SQRT(2)*X + 2*SQRT(2) - 2*

$SQRT(X^4 + 4*X^3 + 2*X^2 + 1)) - 10*LOG(2*X^2 - 1)*X^2 + 5*LOG(2*X^2 - 1$

$) - 4*LOG(2*SQRT(2)*X + 4*SQRT(2) - X^2 - 4*X - 6)*X^2 + 2*LOG(2*$

$SQRT(2)*X + 4*SQRT(2) - X^2 - 4*X - 6) + 10*LOG(4*SQRT(2)*SQRT(X^4$

$+ 4*X^3 + 2*X^2 + 1) - 16*SQRT(2)*X^2 - 8*SQRT(2)*X + 6*SQRT(2) + 19$

$*SQRT(X^4 + 4*X^3 + 2*X^2 + 1) - 29*X^2 - 38*X + 5)*X^2 - 5*LOG(4*SQRT($

$2)*SQRT(X^4 + 4*X^3 + 2*X^2 + 1) - 16*SQRT(2)*X^2 - 8*SQRT(2)*X + 6*$

$SQRT(2) + 19*SQRT(X^4 + 4*X^3 + 2*X^2 + 1) - 29*X^2 - 38*X + 5) + 4*$

$LOG(10*SQRT(2)*SQRT(X^4 + 4*X^3 + 2*X^2 + 1) + 4*SQRT(2)*X^2 + 16*SQRT$

$(2)*X - 2*SQRT(2) + 17*SQRT(X^4 + 4*X^3 + 2*X^2 + 1) - 11*X^2 - 44*X$

$- 39)*X^2 - 2*LOG(10*SQRT(2)*SQRT(X^4 + 4*X^3 + 2*X^2 + 1) + 4*SQRT(2$

$)*X^2 + 16*SQRT(2)*X - 2*SQRT(2) + 17*SQRT(X^4 + 4*X^3 + 2*X^2 + 1) -$

$11*X^2 - 44*X - 39) + 2*LOG(731*SQRT(2)*SQRT(X^4 + 4*X^3 + 2*X^2 + 1)*$

$X + 71492*SQRT(2)*SQRT(X^4 + 4*X^3 + 2*X^2 + 1) - 40957*SQRT(2)*X^3 -$

$174250*SQRT(2)*X^2 - 122871*SQRT(2)*X + 50648*SQRT(2) - 70030*SQRT($

$X^4 + 4*X^3 + 2*X^2 + 1)*X - 141522*SQRT(X^4 + 4*X^3 + 2*X^2 + 1) +$

$90874*X^3 + 403722*X^2 + 272622*X - 61070)*X^2 - LOG(731*SQRT(2)*SQRT$

$(X^4 + 4*X^3 + 2*X^2 + 1)*X + 71492*SQRT(2)*SQRT(X^4 + 4*X^3 + 2*X^2 + 1$

$) - 40957*SQRT(2)*X^3 - 174250*SQRT(2)*X^2 - 122871*SQRT(2)*X +$

$50648*SQRT(2) - 70030*SQRT(X^4 + 4*X^3 + 2*X^2 + 1)*X - 141522*SQRT($

$X^4 + 4*X^3 + 2*X^2 + 1) + 90874*X^3 + 403722*X^2 + 272622*X - 61070)$

$+ 12*LOG((-1))*X^2 - 6*LOG((-1)) + 2*SQRT(X^4 + 4*X^3 + 2*X^2 + 1)*X$

$+ SQRT(X^4 + 4*X^3 + 2*X^2 + 1))/(2*(2*X^2 - 1))$

Example 6: A Nested Expression.

The following example is taken from a well-known table of integrals (Bois, 1961), and so can be said to be more "realistic" than some of the above examples, which were constructed to illustrate particular points. The integrand is $\sqrt{X + \sqrt{A^2 + X^2}}/X$, which involves nested square roots in its current form (which is probably the easiest to manipulate, and is almost certainly more meaningful than a primitive minimal polynomial expression). The multivariate expression for this would be Z/X, where $Z^2 = X + Y$ and $Y^2 = A^2 + X^2$, while the primitive representation for Z is $Z^4 - 2XZ^2 - A^2 = 0$.

The integrand has residues of $\sqrt{A}, i\sqrt{A}, -\sqrt{A}$ and $-i\sqrt{A}$ at the four places lying over 0, and no residues anywhere else. These residues form a 2-dimensional Z-module, so there are two different divisors to consider. In fact both are of order 1, with corresponding functions

$$(-YZ\sqrt{A} + YA - ZA\sqrt{A} + Z\sqrt{A}X + A^2)/X \text{ and}$$
$$(-YZi\sqrt{A} - YA + Zi\sqrt{A}A + Zi\sqrt{A}X + A^2)/X$$

(computed in 3.52 and 3.66 seconds respectively by Coates' Algorithm).

After these two logarithms are subtracted from the integral, the rest of the integrand is

$$\frac{Z(YX + A^2 + X^2)}{Y*A^2 + Y*X^2 + A^2*X + X^3}$$

which integrates to $2Z$. Hence the integral is:

$$\log(-yz\sqrt{a} + ya - za\sqrt{a} + z\sqrt{a}x + a^2) +$$

$$\log(-yzi\sqrt{a} - ya + zi\sqrt{a}a + zi\sqrt{a}x + a^2) - \sqrt{a}(1 + i)\log x + 2z.$$

This expression for the integral is definitely different from that of the standard table (Bois, 1961), since that is expressed in terms of inverse trigonometric forms. More

interestingly, the two forms are not equivalent (even allowing for the possibility of differing constants of integration). In fact the answer in the standard tables is wrong, and is the
integral of Z/Y.

The entire integration process took 12 seconds, and this includes the two Coates'
Algorithm steps and the time required to manipulate all the algebraic expressions and to
decide that they were independent. Furthermore, the version of the program distributed
with REDUCE-2 (i.e. not the one intended for research on the problem of integrating
algebraic expressions) has various heuristic sections in it which often enable it to "guess"
the correct answer without needing to do all the work. This version also has many
"diagnostic" printing statements removed and does not perform as many internal consisten-
cy checks. It is important to note that none of these changes affect the completeness of
the program's output or mean that it can state that something is not integrable when in fact
it is: they merely reflect the differences between a program designed for development and
one designed for production. This program took 1.55 seconds to perform this integral.

Another change that probably ought to be made in a production version is the
insertion of a test for the existence of less "expensive" expressions for algebraics where
these can be found simply. For example, in the example described above we often have to
deal with the square root of A, which is an algebraic expression requiring considerable care
in its handling (see Chapter 2). If we replaced A by B^2 throughout, then this problem
would not arise, and the program could be expected to run faster. In the case above we
could expect to save about 10% of the total time this way.

Example 7: Logarithmically Unintegrable.

Consider the problem of integrating $1/Y*X$ where $Y^2 = X^3 - 3X^2 + X + 1$. The
integrand is a differential of the second kind, and has residues of ± 1 at the two places
lying over $X=0$. In fact the points over $X=0$ are points of infinite order on the elliptic
curve, as the program discovers by trying to apply Coates' algorithm to $D, D^2, ..., D^{10}, D^{12}$ in
turn. Since none of these divisors are principal, we can use Mazur's bound and assert that
no non-zero multiple of the divisor is principal, and hence that the function is unintegrable.

Note that this is different from Example 2 above, where there were no residues. Here we have residues, but cannot find a logarithm which gives rise to them when differentiated.

The integration took 57.4 seconds, of which the 11 Coates steps took .5, 1.1, 1.3, 1.8, 2.3, 3.3, 4.0, 5.4, 6.5, 9.4 and 15.4 seconds respectively. The last application of Coates' algorithm (i.e. to the divisor $12D$) produced the following set of functions as a basis for the space of functions with the correct poles:

$$(6458746112852301356236562338175490838880*SQRT(X^3 - 3*X^2 + X + 1)*X^5$$

$$+ 126138337392550826553734563041284369888*lastsqrt*X^4 -$$

$$1716496198848775432900079488936638638592*lastsqrt*X^3 -$$

$$264366112672565270849916395223433129984*lastsqrt*X^2 +$$

$$105900251808836156512187623923268171 9808*lastsqrt*X +$$

$$38304705431785970833769814248945863 0656*lastsqrt +$$

$$2670880269837154240949218236831038 40001*X^7 +$$

$$4651627736288846824553149353698845 69800*X^6 -$$

$$2253610682756379002381041473931584 384648*X^5 -$$

$$3303095327519116496960400400396180 69408*X^4 +$$

$$3066809088286454744454857763284048 545536*X^3 +$$

$$3573163168949065143377377757152462 544896*X^2 -$$

$$1250526045247291419290725310477411 035136*X -$$

$$38304705431785970833769814248945863065 6)/X^7$$

$$(6458746112852301356236562338175490838 80*SQRT(X^3 - 3*X^2 + X + 1)*X^5$$

$$+ 1261383373925508265537345630412843698 88*lastsqrt*X^4 -$$

$$1716496198848775432900079488936638638592* lastsqrt*X^3 -$$

$$2643661126725652708499163952234331299 84*lastsqrt*X^2 +$$

$$1059002518088361565121876239232681719 808*lastsqrt*X +$$

$$38304705431785970833769814248945863065 6*lastsqrt +$$

$$26708802698371542409492182368310384000 1*X^7 +$$

$$46516277362888468245531493536988456980 0*X^6 -$$

$$22536106827563790023810414739315843846 48*X^5 -$$

$$33030953275191164969604004003961806940 8*X^4 +$$

$$30668090882864547444548577632840485455 36*X^3 +$$

$$35731631689490651433773775715246254489 6*X^2 -$$

$$1250526045247291419290725310477411035 136*X -$$

$$38304705431785970833769814248945863065 6)/X^8$$

$$(6458746112852301356236562338175490838 80*SQRT(X^3 - 3*X^2 + X + 1)*X^5$$

$$+ 1261383373925508265537345630412843698 88*lastsqrt*X^4 -$$

$$1716496198848775432900079488936638638592*lastsqrt*X^3 -$$

$$26436611267256527084991639522343312 9984*lastsqrt*X^2 +$$

$$10590025180883615651218762392326817 19808*lastsqrt*X +$$

$$38304705431785970833769814248945863 0656*lastsqrt +$$

$$26708802698371542409492182368310384 0001*X^7 +$$

$$46516277362888468245531493536988456 9800*X^6 -$$

$$22536106827563790023810414739315843 84648*X^5 -$$

$$33030953275191164969604004003961806 9408*X^4 +$$

$$30668090882864547444548577632840485 45536*X^3 +$$

$$35731631689490651433773775715246254 4896*X^2 -$$

$$12505260452472914192907253104774110 35136*X -$$

$$38304705431785970833769814248945863 0656)/X^9$$

$$(64587461128523013562365623381754908 3880*SQRT(X^3 - 3*X^2 + X + 1)*X^5$$

$$+ 12613833739255082655373456304128436 9888*lastsqrt*X^4 -$$

$$17164961988487754329000794889366386 38592*lastsqrt*X^3 -$$

$$26436611267256527084991639522343312 9984*lastsqrt*X^2 +$$

$$10590025180883615651218762392326817 19808*lastsqrt*X +$$

383047054317859708337698142489458630656*lastsqrt +

$$2670880269837154240949218236831038400001 * X^7 +$$

$$465162773628884682455314935369884569800 * X^6 -$$

$$225361068275637900238104147393158438464 8 * X^5 -$$

$$330309532751911649696040040039618069408 * X^4 +$$

$$306680908828645474445485776328404854553 6 * X^3 +$$

$$357316316894906514337737757152462544896 * X^2 -$$

$$125052604524729141929072531047741103513 6 * X -$$

$$383047054317859708337698142489458630656) / X^{10}$$

$$(6458746112852301356236562338175490838 80 * SQRT(X^3 - 3 * X^2 + X + 1) * X^5$$

$$+ 126138337392550826553734563041284369888 * lastsqrt * X^4 -$$

$$171649619884877543290007948893663863859 2 * lastsqrt * X^3 -$$

$$264366112672565270849916395223433129984 * lastsqrt * X^2 +$$

$$105900251808836156512187623923268171980 8 * lastsqrt * X +$$

383047054317859708337698142489458630656*lastsqrt +

$$2670880269837154240949218236831038400001 * X^7 +$$

$$465162773628884682455314935369884569800 * X^6 -$$

$$225361068275637900238104147393158438464 8 * X^5 -$$

$$330309532751911164969604004003961806940 8 * X^4 +$$

$$306680908828645474445485776328404854553 6 * X^3 +$$

$$357316316894906514337737757152462544896 * X^2 -$$

$$1250526045247291419290725310477411035136 * X -$$

$$3830470543178597083376981424894586306 56) / X^{11}$$

$$(6458746112852301356236562338175490838 80 * SQRT(X^3 - 3 * X^2 + X + 1) * X^5$$

$$+ 1261383373925508265537345630412843698 88 * lastsqrt * X^4 -$$

$$1716496198848775432900079488936638638592 * lastsqrt * X^3 -$$

$$2643661126725652708499163952234331299 84 * lastsqrt * X^2 +$$

$$1059002518088361565121876239232681719808 * lastsqrt * X +$$

$$3830470543178597083376981424894586306 56 * lastsqrt +$$

$$2670880269837154240949218236831038400 01 * X^7 +$$

$$4651627736288846824553149353698845698 00 * X^6 -$$

$$225361068275637900238104147393158438464 8 * X^5 -$$

$$330309532751911164969604004003961806940 8 * X^4 +$$

$$306680908828645474445485776328404854553 6 * X^3 +$$

$$3573163168949065143377377571524625448 96*X^2 \ -$$

$$12505260452472914192907253104774110351 36*X \ -$$

$$38304705431785970833769814248945863065 6)/X^{12}$$

$$(\ -\ 2316186053639990532*SQRT(X^3\ -\ 3*X^2\ +\ X\ +\ 1)*X^5\ -$$

$$35486480838630350280*lastsqrt*X^4\ -\ 34287025200284602624*lastsqrt*$$

$$X^3\ +\ 47330709012277142016*lastsqrt*X^2\ +\ 59834813484366611456*$$

$$lastsqrt*X\ +\ 15684088745312724992*lastsqrt\ -\ 16342827998351920001*$$

$$X^6\ -\ 28462807885868569800*X^5\ +\ 82009784546149012824*X^4\ +$$

$$87267876128018823680*X^3\ -\ 51761471543327269632*X^2\ -$$

$$67676857857022973952*X\ -\ 15684088745312724992)/X^7$$

$$(\ -\ 2316186053639990532*SQRT(X^3\ -\ 3*X^2\ +\ X\ +\ 1)*X^5\ -$$

$$35486480838630350280*lastsqrt*X^4\ -\ 34287025200284602624*lastsqrt*$$

$$X^3\ +\ 47330709012277142016*lastsqrt*X^2\ +\ 59834813484366611456*$$

$$lastsqrt*X\ +\ 15684088745312724992*lastsqrt\ -\ 16342827998351920001*$$

$$X^6\ -\ 28462807885868569800*X^5\ +\ 82009784546149012824*X^4\ +$$

$$87267876128018823680*X^3\ -\ 51761471543327269632*X^2\ -$$

$$67676857857022973952*X\ -\ 15684088745312724992)/X^8$$

$$(- 2316186053639990532*SQRT(X^3 - 3*X^2 + X + 1)*X^5 -$$

$$35486480838630350280*lastsqrt*X^4 - 34287025200284602624*lastsqrt*$$

$$X^3 + 47330709012277142016*lastsqrt*X^2 + 59834813484366611456*$$

$$lastsqrt*X + 15684088745312724992*lastsqrt - 16342827998351920001*$$

$$X^6 - 28462807885868569800*X^5 + 82009784546149012824*X^4 +$$

$$87267876128018823680*X^3 - 51761471543327269632*X^2 -$$

$$67676857857022973952*X - 15684088745312724992)/X^9$$

$$(- 2316186053639990532*SQRT(X^3 - 3*X^2 + X + 1)*X^5 -$$

$$35486480838630350280*lastsqrt*X^4 - 34287025200284602624*lastsqrt*$$

$$X^3 + 47330709012277142016*lastsqrt*X^2 + 59834813484366611456*$$

$$lastsqrt*X + 15684088745312724992*lastsqrt - 16342827998351920001*$$

$$X^6 - 28462807885868569800*X^5 + 82009784546149012824*X^4 +$$

$$87267876128018823680*X^3 - 51761471543327269632*X^2 -$$

$$67676857857022973952*X - 15684088745312724992)/X^{10}$$

$$(- 2316186053639990532*SQRT(X^3 - 3*X^2 + X + 1)*X^5 -$$

$$35486480838630350280*lastsqrt*X^4 - 34287025200284602624*lastsqrt*$$

$$X^3 + 47330709012277142016*lastsqrt*X^2 + 59834813484366611456*$$

$$lastsqrt*X + 15684088745312724992*lastsqrt - 16342827998351920001*$$

$$X^6 - 28462807885868569800*X^5 + 82009784546149012824*X^4 +$$

$$87267876128018823680*X^3 - 51761471543327269632*X^2 -$$

$$67676857857022973952*X - 15684088745312724992)/X^{11}$$

$$(- 2316186053639990532*SQRT(X^3 - 3*X^2 + X + 1)*X^5 -$$

$$35486480838630350280*lastsqrt*X^4 - 34287025200284602624*lastsqrt*$$

$$X^3 + 47330709012277142016*lastsqrt*X^2 + 59834813484366611456*$$

$$lastsqrt*X + 15684088745312724992*lastsqrt - 16342827998351920001*$$

$$X^6 - 28462807885868569800*X^5 + 82009784546149012824*X^4 +$$

$$87267876128018823680*X^3 - 51761471543327269632*X^2 -$$

$$67676857857022973952*X - 15684088745312724992)/X^{12}$$

There is no linear combination of these functions with the appropriate 12-fold zero, so this divisor is indeed not principal.

The size of these expressions, compared with the original integrand, shows how much this algorithm can suffer from "intermediate expression swell". It is worth noting that, if we had used a Lutz-Nagell test (see chapter 7) we would have known immediately that the divisor was not a torsion divisor. This is an efficiency point that should probably be dealt with in the implementation.

Example 8: Sum of two Functions.

Let us consider the integral of $1/f + k/g$, where k is a constant and $f = \sqrt{X^2-1}, g = \sqrt{X^2-4}$. This integrand has residues of $\pm 1 \pm k$ at the 4 places lying over infinity. Therefore if k is not rational, the space of residues is two-dimensional, and the logarithmic part is the sum of two terms, one corresponding to $1/f$ and the other to $1/g$.

When k is rational (say p/q with p,q integers), then the space of residues is one-dimensional, with a basis element of $1/q$ and coefficients of $\pm(p \pm q)$. When p and q are both odd, then $p \pm q$ is always even, and the basis element can be taken as $2/q$. In fact the cases of $1/q$ and $2/q$ are different, and we will treat them separately.

Let us consider a "$1/q$" case, and we will run this both using the algorithm COATES directly, and then using the algorithm DIVISOR__TO__FUNCTION to show how, in certain cases, use of this improved technique can lead to faster runs and more meaningful output.

```
INT(1/SQRT(X**2-1)+10/SQRT(X**2-4),X);
PLACES AT WHICH POLES COULD OCCUR
X=infinity
X=-1
X=1
X=2
X=-2
FIND FUNCTION WITH ZEROS OF ORDER:(-11 9 -9 11)
FIND FUNCTION WITH ZEROS OF ORDER:(-11 -11 -11 -11 20 2 22)
(COATES TIME 42316 MILLISECONDS)
EXTENSION LOGARITHM IS
                  9                    7                    5
  - ANS1*ANS2*X  + 8*ANS1*ANS2*X   - 21*ANS1*ANS2*X   + 20*ANS1*ANS2*
  3                          10              8              6             4
X   - 5*ANS1*ANS2*X - ANS1*X    + 8*ANS1*X    - 21*ANS1*X   + 20*ANS1*X
              2            10            8             6            4
  - 5*ANS1*X   - ANS2*X    + 10*ANS2*X   - 35*ANS2*X   + 50*ANS2*X   - 25
        2              11        9        7        5        3
*ANS2*X   + 2*ANS2 - X    + 10*X   - 35*X   + 50*X   - 25*X   + 2*X
    WHERE
```

$$ANS2 := SQRT(X^2 - 1)$$

$$ANS1 := SQRT(X^2 - 4)$$

INNER WORKING YIELDS

LOG((-1)) + ANS1 + ANS2

   WHERE

$$ANS2 := LOG(SQRT(X^2 - 4)*X^9 - 8*SQRT(X^2 - 4)*X^7 + 21*SQRT(X^2 - 4)*X^5 - 20*SQRT(X^2 - 4)*X^3 + 5*SQRT(X^2 - 4)*X + X^{10} - 10*X^8 + 35*X^6 - 50*X^4 + 25*X^2 - 2)$$

$$ANS1 := LOG(SQRT(X^2 - 1) + X)$$

WITH DERIVATIVE

$$(11*ANS1*ANS2*X^{12} - 122*ANS1*ANS2*X^{10} + 503*ANS1*ANS2*X^8 - 954*ANS1*ANS2*X^6 + 835*ANS1*ANS2*X^4 - 290*ANS1*ANS2*X^2 + 20*ANS1*ANS2 + 11*ANS1*X^{13} - 123*ANS1*X^{11} + 515*ANS1*X^9 - 1007*ANS1*X^7 + 939*ANS1*X^5 - 375*ANS1*X^3 + 40*ANS1*X + 11*ANS2*X^{13} - 144*ANS2*X^{11} + 725*ANS2*X^9 - 1760*ANS2*X^7 + 2115*ANS2*X^5 - 1152*ANS2*X^3 + 208*ANS2*X + 11*X^{14} - 145*X^{12} + 739*X^{10} - 1835*X^8 + 2305*X^6 - 1377*X^4 + 310*X^2 - 8)/(ANS1*ANS2*X^{13} - 13*ANS1*ANS2*X^{11} + 65*ANS1*ANS2*X^9 - 157*ANS1*ANS2*X^7 + 189*ANS1*ANS2*X^5 - 105*ANS1*ANS2*X^3 + 20*ANS1*ANS2*X + ANS1*X^{14} - 13*ANS1*X^{12} + 65*ANS1*X^{10} - 157*ANS1*X^8 + 189*ANS1*X^6 - 105*ANS1*X^4 + 20*ANS1*X^2 + ANS2*X^{14} - 15*ANS2*X^{12} + 89*ANS2*X^{10} - 265*ANS2*X^8 + 415*ANS2*X^6 - 327*ANS2*X^4 + 110*ANS2*X^2 - 8*ANS2 + X^{15} - 15*X^{13} + 89*X^{11} - 265*X^9 + 415*X^7 - 327*X^5 + 110*X^3 - 8*X)$$

   WHERE

$$ANS2 := SQRT(X^2 - 1)$$

$$ANS1 := SQRT(X^2 - 4)$$

```
(TIME TAKEN 47144 MILLISECONDS)
```

$$\text{LOG}(\text{SQRT}(X^2 - 1) + X) + \text{LOG}(\text{SQRT}(X^2 - 4)*X^9 - 8*\text{SQRT}(X^2 - 4)*X^7 +$$

$$21*\text{SQRT}(X^2 - 4)*X^5 - 20*\text{SQRT}(X^2 - 4)*X^3 + 5*\text{SQRT}(X^2 - 4)*X + X^{10} -$$

$$10*X^8 + 35*X^6 - 50*X^4 + 25*X^2 - 2) + \text{LOG}((-1))$$

Now using the algorithm DIVISOR_TO_FUNCTION:

```
INT(1/SQRT(X**2-1)+10/SQRT(X**2-4),X);
PLACES AT WHICH POLES COULD OCCUR
X=infinity
X=-1
X=1
X=2
X=-2
OPERATE ON DIVISOR:(11 9 -9 -11)
AT
X=infinity--
X=infinity+-
X=infinity-+
X=infinity++
FIND FUNCTION WITH ZEROS OF ORDER:(-1 -1 1)
FIND FUNCTION WITH ZEROS OF ORDER:(-1 -1 -1 -1 2 1)
REPLACED BY THE POLE
X=infinity+-
9 TIMES
FIND FUNCTION WITH ZEROS OF ORDER:(2 -2)
FIND FUNCTION WITH ZEROS OF ORDER:(-2 -2 -2 -2 4 2 2)
(COATES TIME 2948 MILLISECONDS)
EXTENSION LOGARITHM IS
ANS1
   WHERE
```

$$\text{ANS1} := \text{SQRT}(X^2 - 1)*\text{SQRT}(X^2 - 4) + \text{SQRT}(X^2 - 1)*X + \text{SQRT}(X^2 - 4)*X + X^2*\text{SQRT}(X^2 - 4) + X^9$$

```
INNER WORKING YIELDS
ANS1 + 10*ANS2
   WHERE
```

$$\text{ANS2} := \text{LOG}(\text{SQRT}(X^2 - 4) + X)$$

$$\text{ANS1} := \text{LOG}(\text{SQRT}(X^2 - 1) + X)$$

```
WITH DERIVATIVE
```

$$(11*ANS1*ANS2*X^3 - 14*ANS1*ANS2*X^4 + 11*ANS1*X - 15*ANS1*X^2 + 4*$$
$$ANS1 + 11*ANS2*X^4 - 54*ANS2*X^2 + 40*ANS2 + 11*X^5 - 55*X^3 + 44*X)/($$
$$ANS1*ANS2*X^4 - 5*ANS1*ANS2*X^2 + 4*ANS1*ANS2 + ANS1*X^5 - 5*ANS1*$$
$$X^3 + 4*ANS1*X + ANS2*X^5 - 5*ANS2*X^3 + 4*ANS2*X + X^6 - 5*X^4 + 4*$$
$$X^2)$$
WHERE

$$ANS2 := SQRT(X^2 - 1)$$

$$ANS1 := SQRT(X^2 - 4)$$
(TIME TAKEN 6107 MILLISECONDS)

$$LOG(SQRT(X^2 - 1) + X) + 10*LOG(SQRT(X^2 - 4) + X)$$

The "$2/q$" case is more complex because the divisor given by the residues is of order 2, i.e. D is not prinicipal, but $2D$ is. However, the analysis of $2D$ is the same as the analysis of D in the case we discussed above, and in particular is much faster with the algorithm DIVISOR-TO-FUNCTION.

Algorithms for Algebraic Expressions

This Appendix contains various algorithms designed for manipulating general algebraic expressions. Many of them can be found, in versions designed for primitive representations rather than multi-variate representations, in the work of Trager (1976).

It is noteworthy that many of them are not guaranteed to work over all fields of finite characteristic, so it is arguable that they are not really algorithms at all. Clearly the area of algebraic algorithms over fields of finite characteristic requires much further study.

SQFR__NORM [*]

Input: K: a field of characteristic p (possibly 0).

A: an algebraic over K, defined by the polynomial P_A.

F(X,A): a square-free polynomial over K(A).

Output: Either FAILED (see footnote) or:

S: a positive integer.

G(X,A): = F(X-SA,A).

R(X): = Norm(G(X,A)) : a square free polynomial over K.

The norm is taken with respect to the field extension K(A)/K.

[1] S := 0.

---

[*] This is taken from the algorithm of the same name by Trager(1976). The difference is that we may wish to apply it when the ground field is of finite characteristic. In this case the algorithm is not guaranteed to succeed, and it may occasionally return the answer FAILED. In this case the computation has to be abandoned, but since this can only occur for finitely many primes, we are not too worried. The reader is referred to Trager's paper for details of the proof of this algorithm.

G(X,A) := F(X,A).

[2] R(X) := Resultant(P_A(Y),G(X,Y),Y).

> The resultant of the minimal polynomial for A and G, taken with respect to
> the dummy variable Y.

[3] If R(X) = SQUARE-FREE-DECOMPOSE(R(X))

> Then return (S,G,R).

[4] S := S+1.

> G(X,A) := G(X-A,A).

[5] If S non-zero

> Then go to [2].
>
> > This can only fail to occur in a field of non-zero characteristic, since S is a
> > positive integer.

[6] Return FAILED.

ALG__FACTOR

Input: F(X): a polynomial over $K(A_1,...,A_n)$.

> K: A purely transcendental field.
>
> $A_1,...,A_n$: A list of algebraic expressions, each A_i having an irreducible minimal
> polynomial P_i over $K(A_1,...,A_{i-1})$.

Output: A list of the factors of F with their multiplicities.

> If called with a field of finite characteristic, it may return the answer
> FAILED indicating that one of the internal sub-algorithms was unable to
> operate over that finite field.

[1] L := SQUARE__FREE__DECOMPOSE(F).

[2] ANSWER := NIL.

[3] For I = 1,...,Length L Do:

[3.1] A := ALG__FACTOR__2(L[I],K,A$_1$,...,A$_n$).

[3.2] If A = FAILED

      Then return FAILED.

[3.3] For each B in A Do:

[3.3.1] ANSWER := (B ** I) . ANSWER.

      The exponentiation here is purely symbolic, indicating that the factor B

      occurs in F with multiplicity I.

[4] Return ANSWER.

ALG__FACTOR__2 [#]

Input: F(X): a square free polynomial over K(A$_1$,...,A$_n$).

      K: A purely transcendental field.

      A$_1$,...,A$_n$: A list of algebraic expressions, each A$_i$ having an irreducible minimal

          polynomial P$_i$ over K(A$_1$,...,A$_{i-1}$).

Output: A list of the factors of F.

[1] (S,G,R) := SQFR__NORM(K(A$_1$,...,A$_{n-1}$),A$_n$,F).

[#] This is similar to the algorithm ALG-FACTOR of Trager (1976) except that it is
intended to work over multivariate representation of algebraic extensions, and hence is
recursive. It calls the algorithm FACTOR, which is intended to factorise its input over a
non-algebraic ground field of any degree or characteristic. Such algorithms are available
by *p*-adic methods (Wang, 1976 or 1978).

It is possible (but only if K is of finite characteristic) for this step to return FAILED, in which case this algorithm has also got to return FAILED.

[2] If n = 1

    Then L := FACTOR(R(X))

    Else L := ALG__FACTOR__2 $(R(X),K,A_1,...,A_{n-1})$.

        This is the recursive step in the factorisation process: FACTOR is used when there is no algebraic dependence to solve[*].

[3] If length(L) = 1

    Then return F.

[4] For each H in L do:

[4.1] $H(X,A_n) := gcd(H(X),G(X,A_n))$.

[4.2] $G(X,A_n) := G(X,A_n)/H(X,A_n)$.

[4.3] $H(X,A_n) := H(X+SA_n,A_n)$.

[5] Return L.

PRIMITIVE__ELEMENT [#]

Input: K: a field.

---

[*] There is a possible optimisation here in the case when K has characteristic 0 and all the A_i are algebraic integers: we could use an algorithm for factorisation over algebraic number fields, e.g. that due to Wang & Rothschild(1975).

[#] This algorithm is similar to the algorithm PRIMITIVE-ELEMENT of Trager (1976) except that it is designed to work over multi-variate representations, and it also returns the expression for the chosen primitive element in terms of the original representation. It will also work over fields of finite characteristic provided that the algorithm SQFR-NORM described above works, which can only fail to happen for finitely many primes.

$A_1,...,A_n$: a set of elements of the algebraic closure of K, with A_i algebraic over $K(A_1,...,A_{i-1})$ defined by the polynomial $P_i(X)$.

Output: R(X): a minimal polynomial for X a primitive element for the field $K(A_1,...,A_n)$ over K

X: the representation of that element as a member of $K(A_1,...,A_n)$ - i.e. a multivariate representation.

$Q_i(X)$: the representations of A_i in terms of the minimal element. These are therefore rational functions in X over K.

[1] If n = 1

then return $(P_1(X),A_1,(P_1(X)))$.

[2] $(R',X',(Q_i'(X'))) := PRIMITIVE\_ELEMENT(K,A_1,...,A_{n-1})$.

This gives us a primitive representation for the field generated by $A_1,...,A_{n-1}$. These answers will not do directly as components of the final answer, because, in adding A_n, the primitive element X will change, and so the polynomials $Q_i(X)$ will need to be changed.

If the characteristic of K is non-zero, then this step could return FAILED. In this case the answer to the entire computation is FAILED.

[3] $(S,G,R) := SQFR-NORM(K,X',P_n)$.

Note that, if our multi-variate representation involved nested expressions, then P_n might involve some of the previous A_i. In this case they must be replaced by their representation in terms of X', viz. the polynomials $P_i'(X')$.

We also note that this step could return FAILED if the characteristic of K is non-zero, and if this happens then we cannot proceed with this computation (even though there still is a primitive element).

[4] $Z := B/A$, where $AX'-B = gcd(G(X',X),R'(X'))$.

This computation is being performed in $K(X)$, where X is a root of the polynomial R, and X' is being treated as the gcd variable.

Z is the representation for X' in terms of X.

[5] $Q_n(X) := X - S*Z$.

This reflects the fact that $X = A_n + S*X'$.

[6] For $i = 1,...,n-1$ do :

[6.1] $Q_i(X) := \text{Substitute}(Q_i'(X'),X' => Z)$.

This loop creates all the defining polynomials for the elements A_i in terms of X.

[7] Return $(R(X),A_n+S*X',(Q_i))$.

SQUARE_FREE_DECOMPOSE [*]

Input: $R(X)$: a polynomial over some ground field K.

Output: $(R_1(X),...,R_n(X))$

where

$$R(X) = \prod_{i=1}^{i=n} R_i(X)^i$$

---

[*] This algorithm, for performing square-free decomposition of polynomials, is almost certainly not original. The only feature of it that I have not been able to find in elementary texts is its treatment of exact p-th powers modulo the prime p. I am sure that I cannot have been the first to invent this, but I do not know of another description.

This is not the most efficient algorithm for use over fields of characteristic 0 (for which see Yun, 1977b), but the more efficient one does not generalise to fields of prime characteristic as readily.

and the R_i are coprime and square-free.

[1] If degree(R) = 1, then return R.

[2] R':= dR/dX

as a formal polynomial derivative.

[3] If R'=0 then

This can only happen when every term in R is an exact p-th power, where p is the characteristic of K, because R has positive degree. In particular this cannot occur when K has characteristic 0.

[3.1] R":= $(R(X))^{1/p}$.

This is achieved by taking $R'' = \sum_{i=0}^{i=k} a_i X^i$, where $R = \sum_{i=0}^{i=k} a_i X^{ip}$ and p is the characteristic of K.

[3.2] L:= SQUARE_FREE_DECOMPOSE(R").

[3.3] Return L'

where L' is L with $(p-1)$ zeros inserted before each element of L, to represent the fact the we want the decomposition of $R = R'^p$.

[4] R" := gcd(R,R').

[5] If R" in K

Then return R.

Then R is square-free, and we just return it.

[6.1] M[1] := (R/R")/gcd(R",R/R").

The answer will be accumulated in the vector M. We observe that this element, although it is computed as a quotient, is in fact a polynomial in X over K.

[6.2] R := R/M[1].

We take out of R each factor as it is discovered.

[7] L:= SQUARE__FREE__DECOMPOSE(R").

[8] For I = 2,... Do:

      Until the non-zero elements of L are exhausted.

[8.1] If p | I

    Then M[I] := L[I]

    Else M[I+1] := L[I].

[8.2] If p | I

    Then R := R/L[I]**I

    Else R := R/L[I]**(I+1).

[9] If R in K

    Then Return M.

[10] For I = p,2p,... Do:

      This loop, which only occurs in the case of non-zero characteristic, is to solve the embarrassment caused by confusing factors of Y^p and Y^{p+1}.

[10.1] R' := gcd(R,M[I]).

[10.2] M[I+1] := R'.

[10.3] M[I] := M[I]/R'.

[10.4] R := R/R'.

[11] Return M.

Lemma 1 The algorithm given above terminates.

Proof: The only way in which it could fail to terminate would be if it recursed indefinitely. But if it calls itself via [3.2], it has decreased the degree of its argument by a factor of p, and if it calls itself via step 7, we have $\deg(R") = \deg(\gcd(R,R')) \leq \deg(R')$ $< \deg(R)$, so the degree is a strictly decreasing integer function. Hence the recursion is finite and the algorithm terminates.

Theorem 2 The Algorithm returns the correct answer.

Proof: Let $R(X) = \prod_{i=1}^{n} R_i(X)^i$. We can assume that $R(X)$ has degree >1, else the problem is trivial. We can now use the previous Lemma to proceed inductively on the degree of R. If $R(X)$ is an exact p-th power (where p is the characteristic of the ground field), then indeed $R'(X) = 0$, and the algorithm proceeds by step 3. Then $R"$ is the p-th root of R and the algorithm is correct.

If R is not an exact p-th power, then

$$R'(X) = \sum_{i=1}^{n} iR_i(X)' \frac{R(X)}{R_i(X)},$$

where $R_i(X)'$ is the derivative of $R_i(X)$. Let $S = \prod_{i=1}^{n} R_i(X)^{i-1}$, $T = \prod_{i=1}^{n} R_i(X)$. Then $R" = \gcd(R,R') =$

$$S \; \gcd(T, \sum_{i=1}^{i=n} iR_i(X)' \frac{T}{R_i(X)}).$$

Since the R_i are coprime, we deduce that the only possible factors of the last gcd above are those R_i for which $i = 0$. Hence $R" = \prod_{i=1}^{n} R_i(X)^{i'}$ where $i' = \begin{cases} i & \text{if } i\,|\,p \\ i-1 & \text{otherwise} \end{cases}$. $R/R"$ is then $\prod_{i=1}^{n} R_i(X)$ (p not dividing i) , and M[1] is then $R_1(X)$.

we can then perform a square-free decomposition of $R"$, which is indeed what the recursive step [7] does. Unfortunately terms of the form $R_{kp}(X)^{kp}$ and $R_{kp+1}(X)^{kp+1}$ R have been confused in $R"$, since they now both have exponent kp. In the decomposition of $R"$, they will both be returned with exponent kp, and the loop at [10] will perform the necessary discrimination, because there will be a residual factor of $R_{kp+1}(X)$ but not of $R_{kp}(X)$. This proves the other path through the algorithm, and hence the entire algorithm.

Bibliography

Abbreviations.

Proc. SYMSAC 76.

      Proceedings of the 1976 ACM Symposium on Symbolic and Algebraic Computation. ACM Inc., New York, 1976.

Proc. EUROSAM 79. Proceedings of the 1979 European Symposium on Symbolic and Algebraic Computation. Springer-Verlag Lecture Notes in Computer Science 72, Berlin-Heidelberg-New York 1979.

Proc. 1977 MACSYMA Users' Conference. NASA publication CP-2012, National Technical Information Service, Springfield, Virginia.

(Baker,1975) Baker,A., Transcendental Number Theory. Cambridge University Press, 1975.

(Baker & Coates,1970) Baker,A. & Coates,J., Integer Points on Curves of genus 1. Proc. Cam. Phil. Soc. 67(1970) pp. 595 et seq.

(Baldassari & Dwork,1979) Baldassari,R., & Dwork,B., On second order linear differential equations with algebraic solutions. Amer. J. Math. 101(1979) pp. 42-76.

(Berlekamp,1967) Berlekamp,E.R., Factoring Polynomials over Finite Fields. Bell System Tech. J. 46(1967) pp. 1853-1859.

(Berlekamp,1970) Berlekamp,E.R., Factoring Polynomials over Large Finite Fields. Math. Comp. 24 (1970) pp. 713-735.

(Birch & Swinnerton-Dyer,1963) Birch,B.J. & Swinnerton-Dyer, H.P.F., Notes on Elliptic Curves I, J. Reine u. Angew. Math. 212(1963) pp. 7-23.

(Bogen et al,1977) Bogen,R.A, et al. MACSYMA Reference Manual. MIT Laboratory for Computer Science, Cambridge, Mass.

(Bois,1961) Bois, G. Petit, A Table of Indefinite Integrals. Dover 1861.

(Borevich & Shafarevich,1966) Borevich,Z.I. & Shafarevich,I.R., Number Theory. Academic Press, New York, 1966 (Translated from Teoria Chisel, Moscow, 1964).

(Buchberger,1979) Buchberger,B. A Criterion for Detecting Unnecessary Reductions in the Construction of Groebner Bases. Proc. EUROSAM 79 pp. 3-21.

(Carlson,1965) Carlson,B.C., On Computing Elliptic Integrals & Functions. J. Math. Phys. 44(1965) pp. 36-51.

(Cassels,1966) Cassels,J.W.S., Diophantine Equations with Special Reference to Elliptic Curves. J. L.M.S. 41(1966) pp. 193-291.

(Caviness,1978) Caviness,B.F., Private Communication, July 1978.

(Caviness & Fateman,1976) Caviness,B.F. & Fateman,R.J., Simplification of Radical Expressions. Proc SYMSAC 76, pp. 329-338.

(Cayley,1853) Cayley,A., Note on the Porism of the in-and-circumscribed Polygon. Philosophical Magazine (4th. ser) VI(1853) pp. 99-103 & 376-7.

(Cayley,1861) Cayley,A., On the Porism of the in-and-circumscribed Polygon. Phil. Trans. Roy. Soc. CLI(1861) pp. 225-239.

(Chebyshev,1821-1894) Chebyshev(Tchebichef),P.L., Oeuvres de. 2 vols., St Pétersbourg. Reprinted Chelsea Publishing Co., New York.

(Chebyshev,1853) Chebyshev(Tchebichef),P.L., Sur l'intégration des différentielles irrationelles. Journal de Maths. Pures et Appls. XVIII(1853) pp. 87-111. Oeuvres (vide supra) vol. I pp. 147-168.

(Chebyshev,1857) Chebyshev(Tchebichef),P.L., Sur l'intégration des différentielles qui contiennent une racine carrée d'un polynôme du troisième ou du quatrième degré. Journal de Maths. Pures et Appl. (2nd. ser) II(1857) pp. 1-42 (taken from Mémoires de l'Academie Impériale des Sciences de Saint-Pétersbourg (6th. ser) Sciences Math. et Phys., vol. VI pp. 203-232). Oeuvres (vide supra) vol. I pp. 171-200.

(Chebyshev,1860) Chebyshev(Tchebichef),P.L., Sur l'intégration de la différentielle $(x + A)dx/\sqrt{x^4 + ax^3 + bx^2 + cx + d}$. Bulletin   de   l'Académie   Impériale   de Saint-Pétersbourg III(1861) pp. 1-12. (Summarised in Comptes Rendus de l'Académie des Sciences LI(1860) p.46-48) (Reprinted with Summary in Journal de Maths. Pures et Appl. (2nd. ser) 9(1864) pp. 225-246). Oeuvres (vide supra) vol. I pp. 517-530.

(Chevalley,1951) Chevalley,C., Introduction to the Theory of Algebraic Functions of one Variable, A.M.S. Surveys VI, 1951.

(Churchhouse,1976) Churchhouse,R.F., Efficient Computation of Algebraic Continued Fractions. Astérisque 38-39 (1976) pp. 23-32.

(Coates,1970) Coates,J., Construction of Rational Functions on a Curve. Proc. Cam. Phil. Soc. 68(1970) pp. 105-123.

(Cohen & Yun,1979) Cohen,J.D., & Yun,D.Y.Y., Algebraic Extensions of Arbitrary Integral Domains. Proc EUROSAM 79 pp. 134-139.

(Collins,1971) Collins,G.E., The Calculation of Multivariate Polynomial Resultants. JACM 18(1971) pp. 515-532.

(Davenport,1979a) Davenport,J.H., The Computerisation of Algebraic Geometry. Proc. EUROSAM 79 pp. 119-133.

(Davenport,1979b) Davenport,J.H., Algorithms for the Integration of Algebraic Functions. Proc EUROSAM 79 pp. 415-425.

(Davenport,1979c) Davenport,J.H., Anatomy of an Integral. SIGSAM Bulletin, November 1979.

(Davenport & Jenks,1980) Davenport,J.H. & Jenks,R.D., MODLISP - an Introduction. Proc. LISP80, The LISP Company, Stanford, California, 1980.

(Demjanenko,1971) Demjanenko,V.A., On the Torsion of Elliptic Curves. Izv. Akad. Nauk. SSSR Ser. Mat. 35(1971) pp. 280-307 (MR 44(1972) #2755).

(Eichler,1966) Eichler,M., Introduction to the Theory of Algebraic Numbers and Functions. Academic Press, London, 1966.

(Epstein,1979) Epstein,H.I., A Natural Structure Theorem for Complex Fields. SIAM J. Computing 8(1979) pp. 320-325.

(ffitch & Norman,1977) ffitch,J.P. & Norman,A.C., Implementing LISP in a High-level Language. Software - Practice and Experience, 7(1977) pp. 713-725.

(Fulton,1969) Fulton,W., Algebraic Curves, An Introduction to Algebraic Geometry. W.A. Benjamin Inc, 1969.

(Griesmer et al.,1975) Griesmer,J.H., Jenks,R.D. & Yun,D.Y.Y., SCRATCHPAD User's Manual. IBM Research Publication RA70, June 1975.

(Griffiths & Harris,1978) Griffiths,P. & Harris,J., On Cayley's Explicit Solution to Poncelet's Porism. L'Enseignement Mathématique (2nd. Series) 24(1974) pp. 31-40.

(Halphen,1886) Halphen,G.H., Traité des fonctions elliptiques et de leurs applications. Paris, 1886-1891.

(Hardy,1916) Hardy,G.H., The Integration of Functions of a Single Variable (2nd. ed.). Cambridge Tract 2, C.U.P.,1916.

(Harrington,1977) Harrington,S.J., A new Symbolic Integration System in Reduce. Utah Computational Physics Report 57, University of Utah, Nov. 1977 (revised May 1978).

(Harrington,1979a) Harrington,S.J., A Symbolic Limit Evaluation Program in REDUCE. SIGSAM Bulletin 49 (Feb. 1979) pp.27-31.

(Harrington,1979b) Harrington,S.J., A new Symbolic Integration System in REDUCE. Computer Journal 22(1979) 2 pp. 127-131.

(Hearn,1973) Hearn,A.C., REDUCE-2 User's Manual.   Computing Physics Group, University of Utah, 1973

(Hearn,1976) Hearn,A.C., A New Reduce Model for Algebraic Simplification.   Proc. SYMSAC 76, pp.46-51.

(Hearn,1979) Hearn,A.C., Non-Modular Computation of Polynomial Gcd Using Trial Division. Proc. EUROSAM 79 pp.227-239.

(Jenks,1979) Jenks.R.D., MODLISP. Proc. EUROSAM 79 pp. 466-480.

(Kaplansky,1957) Kaplansky,I., An Introduction to Differential Algebra, Publications de l'Institut de Mathématique de l'Université de Nancago V, Hermann, Paris, 1957 (2nd. ed. 1976).

(Kenku,1979) Kenku,M.A., Certain Torsion Points on Elliptic Curves defined over Quadratic Fields. J. L.M.S. 2nd. ser. 19(1979) pp. 233-240.

(Kolchin,1973) Kolchin,E.R., Differential Algebra and Algebraic Groups, Academic Press, London, 1973.

(Lang,1957) Lang,S. Introduction to Algebraic Geometry. Interscience, New York, 1957.

(Lang,1959) Lang,S., Abelian Varieties. Interscience, New York, 1959.

(Lang,1960) Lang,S., Integral Points on Curves, Publ. Math. IHES 6(1960), pp. 319-335.

(Lang,1972) Lang,S., Introduction to Algebraic and Abelian Functions. Addison-Wesley, 1972.

(Lang,1978) Lang,S., Elliptic Curves Diophantine Analysis. Springer-Verlag, Berlin-Heidelberg-New York, 1978.

(Lang & Neron,1959) Lang,S. & Neron,A., Rational Points of Abelian Varieties over Function Fields. American J. of Math. 81(1959) pp. 95-118.

(Laplace,1820) Laplace,(P.S.) Marquis de, Théorie Analytique des Probabilités, 3rd. ed., Courcier, Paris, 1820. In: Laplace, Oeuvres complètes du Marquis de, Gauthier-Villars, Paris, 1886, vol. 7 (of 14); or Laplace, Oeuvres de, Imprimerie Royale, Paris, 1847, vol. 7 (of 7).

(Liouville,1833a) Liouville,J., Premier Mémoire sur la Détermination des Intégrales dont la Valeur est Algébrique. Journal de l'Ecole Polytechnique 14(1833) cahier 22, pp. 124-148

(Liouville,1833b) Liouville,J., Second Mémoire sur la Detérmination des Intégrales dont la Valeur est Algébrique. Journal de l'Ecole Polytechnique 14(1833) cahier 22, pp. 149-193

(Liouville,1833c) Liouville,J., Mémoire sur les Transcendentes Elliptiques de Première et Seconde Espèce, Considerées comme Fonctions de leur Amplitude. Journal de l'Ecole Polytechnique 14(1833) cahier 23, pp. 57-83.

(Lipson,1969) Lipson,J.D., Symbolic Methods for the Computer Solution of Linear Equations with Applications to Flow-Graphs. Proc. 1968 Summer Institute on Symbolic Mathematical Computation. IBM, Yorktown Heights, 1969, pp. 233-303.

(Manin,1958) Manin,Ju.I., Algebraic Curves over Fields with Differentiation. Izv. Akad. Nauk. SSSR Ser. Mat. 22(1958) pp. 737-756 (translated in AMS Trans. Ser. 2 37(1964) pp. 59-78).

(Manin,1963) Manin,Ju.I., Rational Points of Algebraic Curves over Function Fields. Izv. Akad. Nauk. SSSR Ser. Mat. 27(1963) pp. 1395-1440 (translated in AMS Trans. Ser. 2 50(1966) pp. 189-234).

(Manin,1969) Manin,Ju.I., Uniform bounds for p-torsion on elliptic curves. Izv. Akad. Nauk. SSSR 33(1969) pp. 459-465. (MR 42 #7667.) See also Serre,1971.

(Mazur,1977) Mazur,B., Rational Points on Modular Curves, in Modular Functions of One Variable V, Springer Lecture Notes in Mathematics 601, Berlin-Heidelberg-New York 1977 (Proceedings International Conference on Modular Functions, Bonn 1976) pp. 107-148.

(Mazur,1978) Mazur,B., Rational Isogenies of Prime Degree. Inventiones Math. 44(1978) pp. 129-162.

(Mazur & Swinnerton-Dyer,1974) Mazur,B. & Swinnerton-Dyer, H.P.F., Arithmetic of Weil Curves. Inventiones Math. 25 (1974) pp. 1-61.

(Mignotte,1976) Mignotte,M., Factorisation des Polynômes sur un Corps Fini. Astérisque 38-39 (1976) pp. 149-157.

(Mordell,1922) Mordell,L.J., On the Rational Solutions of the Indeterminate Equations of the Third and Fourth Degrees. Proc. Cam. Phil. Soc. 21(1922) pp. 179-192.

(Moses,1967) Moses,J., Symbolic Integration. Project MAC report 47, M.I.T., 1967.

(Moses,1971) Moses,J., Symbolic Integration, the stormy decade. Communications ACM 14(1971) pp. 548-560.

(Mumford,1965) Mumford,D., Geometric Invariant Theory. Springer-Verlag, Berlin-Heidelberg-New York, 1965.

(Ng,1974) Ng,E.W., Symbolic Integration of a Class of Algebraic Functions. NASA Technical Memorandum 33-713 (1974).

(Norman,1975) Norman,A.C., Computing with Formal Power Series. ACM Transactions on Mathematical Software 1(1975) pp. 346-356.

(Norman,1978a) Norman,A.C., Symbolic and Algebraic Modes in REDUCE. REDUCE Newsletter 3 (July 1978) pp. 5-9.

(Norman,1978b) Norman,A.C., Towards a REDUCE Solution to SIGSAM Problem 7. SISGAM Bulletin 48 (Nov. 1978) pp. 14-18.

(Norman & Davenport,1979) Norman,A.C., & Davenport,J.H., Symbolic Integration - the Dust Settles? Proc. EUROSAM 79 pp. 398-407.

(Norman & Moore,1977) Norman,A.C. and Moore,P.M.A., Implementing the New Risch Integration Algorithm. Proc. 4th. Int. Colloquium on Advanced Computing Methods in Theoretical Physics, Marseilles, 1977.

(Richardson,1968) Richardson,D., Some Unsolvable Problems Involving Elementary Functions of a Real Variable. Journal of Symbolic Logic 33(1968), pp. 511-520.

(Risch,1969) Risch,R.H., The Problem of Integration in Finite Terms. Trans. A.M.S. 139(1969) pp. 167-189 (MR 38 #5759).

(Risch,1970) Risch,R.H., The Solution of the Problem of Integration in Finite Terms. Bulletin AMS 76(1970) pp. 605-608.

(Risch,1974) Risch,R.H., A Generalization and Geometric Interpretation of Liouville's Theorem on Integration in Finite Terms. IBM Research Report RC 4834 (6 May 1974).

(Ritt,1948) Ritt,J.F., Integration in Finite Terms, Liouville's Theory of Elementary Methods. Columbia University Press, New York, 1948.

(Ritt,1950) Ritt,J.F., Differential Algebra. American Mathematical Society Colloquium Proceedings vol. XXXIII, Providence R.I., 1950 (reprinted Dover, New York, 1966).

(Rosenlicht,1976) Rosenlicht,M., On Liouville's Theory of Elementary Functions. Pacific J. Math 65(1976), pp. 485-492.

(Rothstein,1977) Rothstein,M., A New Algorithm for the Integration of Exponential and Logarithmic Functions. Proc. 1977 MACSYMA Users' Conference, pp. 263-274.

(Rothstein & Caviness,1979) Rothstein,M. & Caviness,B.F., A Structure Theorem for Exponential and Primitive Functions. SIAM J. Computing (to appear).

(Schinzel,1962) Schinzel,A., On Some Problems of the Arithmetical Theory oF Continued fractions II. Acta Arithmetica VII (1962) pp. 287-298.

(Seidenberg,1968) Seidenberg,A., Elements of the Theory of Algebraic Curves. Addison-Wesley, 1968

(Serre,1971) Serre,J.-P., p-torsion des courbes elliptiques (d'après Y. Manin). Séminaire Bourbaki 69/70 no. 380. Springer Lecture Notes in Mathematics 180 (Berlin-Heidelberg-New York, 1971).

(Serre & Tate,1968) Serre,J.-P. & Tate,J.T., Good Reduction of Abelian Varieties. Annals of Mathematics 88 (1968) pp. 492-517.

(Shafarevich,1972) Shafarevich,I.R., Osnovy Algebraicheskoi Geometrii, Nauka, Moscow, 1972. English translation: Basic Algebraic Geometry, Springer-Verlag, Berlin-Heidelberg-New York,1974.

(Shimura & Taniyama,1961) Shimura,G. & Taniyama,A. Complex Multiplication of Abelian Varieties and its Applications to Number Theory. Publ. Math. Soc. Japan 6(1961).

(Shtokhamer,1977) Shtokhamer,R., Attempts in Local Simplification of Non-Nested Radicals. SIGSAM Bulletin 41 (Feb. 1977) pp. 20-21.

(Siegel,1929) Siegel,C.L., Ueber einige Anwendungen Diophantischer Approximationen. Abh. Preuss. Akad. Wiss. (1929) pp. 41-69.

(Siegel,1969) Siegel,C.L., Abschaetzung von Einheiten. Nachr. Akad. Wiss. Goettingen (1969) pp. 71-86.

(Slagle,1961) Slagle,J., A Heuristic Program that Solves Symbolic Integration Problems in Freshman Calculus. Ph.D. Dissertation, Harvard U., Cambridge, Mass. May 1961.

(Smit,1976) Smit,J., The Efficient Calculation of Symbolic Determinants. Proc SYMSAC 76, pp. 105-113.

(Smit,1979) Smit,J., New Recursive Minor Expansion Algorithms, a Presentation in a Comparative Context. Proc. EUROSAM 79, pp. 74-87

(Stephens,1970) Stephens,N.M., Notes on the Algorithm of Birch & Swinnerton-Dyer. Unpublished, 1970.

(Stoutemyer,1977) Stoutemyer,D.R., $\sin(x)**2 + \cos(x)**2 = 1$. Proc 1977 MACSYMA Users' Conference, pp. 425-433.

(Swinnerton-Dyer,1975) Swinnerton-Dyer,H.P.F., Numerical Tables on Elliptic Curves. In: Modular Functions of One Variable IV (Proceedings Antwerp 1972), Springer Lecture Notes in Mathematics 476(Berlin-Heidelberg-New York, 1975).

(Tate,1974) Tate,J.T., The Arithmetic of Elliptic Curves. Inventiones Math. 23(1974) pp. 179-206.

Tchebichef - see Chebyshev.

(Trager,1976) Trager,B.M., Algebraic Factoring and Rational Function Integration. Proc. SYMSAC 76, pp. 219-226.

(Trager,1978) Trager,B.M., IBM Yorktown Heights Integration Workshop, 28-9 Aug. 1978. (Tape recording available from Dr. D.Y.Y. Yun)

(Trager,1979) Trager,B.M., Integration of Simple Radical Extensions. Proc. EUROSAM 79 pp. 408-414.

(van der Waerden,1949) van der Waerden,B.L., Modern Algebra, Frederick Ungar, New York, 1949 (Translated from Moderne Algebra 2nd. ed.).

(Wang,1976) Wang,P.S., Factoring Multivariate Polynomials over Algebraic Number Fields. Math. Comp. 30(1976) pp. 324-336.

(Wang,1978) Wang,P.S., An Improved Multivariable Polynomial Factorising Algorithm. Math. Comp. 32(1978) pp. 1215-1231.

(Wang & Minamikawa,1976) Wang,P.S. & Minamikawa,T., Taking Advantage of Zero Entries in the Exact Inverse of Sparse Matrices. Proc. SYMSAC 76, pp. 346-350.

(Wang & Rothschild,1975) Wang,P.S. & Rothschild,L.P., Factoring Multi-Variate Polynomials over the Integers. Math. Comp. 29(1975) pp. 935-950.

(Weil,1928) Weil,A., L'Arithmétique sur les Courbes Algébriques. Acta. Math. 52(1928) pp. 281-315.

(Wienberger & Rothschild,1976) Factoring Polynomials over Algebraic Number Fields. ACM Transactions on Mathematical Software 2(1976) pp. 335-350.

(Weiss,1963) Weiss,E., Algebraic Number Theory. McGraw-Hill, New York, 1963.

(Whittaker & Watson,1927) Whittaker,E.T., & Watson,G.N., A Course of Modern Analysis. C.U.P. 4th. ed 1927.

(Yun,1973) Yun,D.Y.Y., The Hensel Lemma in Algebraic Manipulation. M.I.T. Thesis MAC TR-138, 1973.

(Yun,1976) Yun,D.Y.Y., Algebraic Algorithms using p-adic Techniques. Proc. SYMSAC 76, pp. 248-259.

(Yun,1977a) Yun,D.Y.Y., Fast Algorithm for Rational Function Integration. IBM Research Report RC 6563 (6 Jan. 1977). Proceedings IFIP 77, Toronto, Canada.

(Yun,1977b) Yun,D.Y.Y., On the Equivalence of Polynomial Gcd and and Squarefree Factorization Algorithms. Proc. 1977 MACSYMA Users' Conference, pp. 65-70.

(Yun & Gustavson,1979) Yun,D.Y.Y., & Gustavson,F., Fast Computation of the Rational Hermite Interpolant and Solving Toeplitz Systems of Equations via the Extended Euclidean Algorithm. Proc. EUROSAM 79, pp. 58-65.

(Zimmer,1972) Zimmer,H.G., Computational Problems, Methods, and Results in Algebraic Number Theory. Lecture Notes in Mathematics 262, Springer-Verlag, Berlin-Heidelberg-New York, 1972.

(Zimmer et al.,1979) Bartz,H., Fischer,K., Folz,H. and Zimmer,H.G., Some Computations relating to Torsion Points of Elliptic Curves over Algebraic Number Fields. Proc EUROSAM 79 pp. 108-118.

(Zippel,1977) Zippel, R.E.B., Radical Simplification Made Easy, Proc. 1977 MACSYMA Users' Conference, pp. 361-367.

(Zippel,1979) Zippel,R.E.B., Probabilistic Algorithms for Sparse Polynomials. Proc. EUROSAM 79, pp. 216-226.

Vol. 49: Interactive Systems. Proceedings 1976. Edited by A. Blaser and C. Hackl. VI, 380 pages. 1976.

Vol. 50: A. C. Hartmann, A Concurrent Pascal Compiler for Mini-computers. VI, 119 pages. 1977.

Vol. 51: B. S. Garbow, Matrix Eigensystem Routines – Eispack Guide Extension. VIII, 343 pages. 1977.

Vol. 52: Automata, Languages and Programming. Fourth Colloquium, University of Turku, July 1977. Edited by A. Salomaa and M. Steinby. X, 569 pages. 1977.

Vol. 53: Mathematical Foundations of Computer Science. Proceedings 1977. Edited by J. Gruska. XII, 608 pages. 1977.

Vol. 54: Design and Implementation of Programming Languages. Proceedings 1976. Edited by J. H. Williams and D. A. Fisher. X, 496 pages. 1977.

Vol. 55: A. Gerbier, Mes premières constructions de programmes. XII, 256 pages. 1977.

Vol. 56: Fundamentals of Computation Theory. Proceedings 1977. Edited by M. Karpiński. XII, 542 pages. 1977.

Vol. 57: Portability of Numerical Software. Proceedings 1976. Edited by W. Cowell. VIII, 539 pages. 1977.

Vol. 58: M. J. O'Donnell, Computing in Systems Described by Equations. XIV, 111 pages. 1977.

Vol. 59: E. Hill, Jr., A Comparative Study of Very Large Data Bases. X, 140 pages. 1978.

Vol. 60: Operating Systems, An Advanced Course. Edited by R. Bayer, R. M. Graham, and G. Seegmüller. X, 593 pages. 1978.

Vol. 61: The Vienna Development Method: The Meta-Language. Edited by D. Bjørner and C. B. Jones. XVIII, 382 pages. 1978.

Vol. 62: Automata, Languages and Programming. Proceedings 1978. Edited by G. Ausiello and C. Böhm. VIII, 508 pages. 1978.

Vol. 63: Natural Language Communication with Computers. Edited by Leonard Bolc. VI, 292 pages. 1978.

Vol. 64: Mathematical Foundations of Computer Science. Proceedings 1978. Edited by J. Winkowski. X, 551 pages. 1978.

Vol. 65: Information Systems Methodology, Proceedings, 1978. Edited by G. Bracchi and P. C. Lockemann. XII, 696 pages. 1978.

Vol. 66: N. D. Jones and S. S. Muchnick, TEMPO: A Unified Treatment of Binding Time and Parameter Passing Concepts in Programming Languages. IX, 118 pages. 1978.

Vol. 67: Theoretical Computer Science, 4th GI Conference, Aachen, March 1979. Edited by K. Weihrauch. VII, 324 pages. 1979.

Vol. 68: D. Harel, First-Order Dynamic Logic. X, 133 pages. 1979.

Vol. 69: Program Construction. International Summer School. Edited by F. L. Bauer and M. Broy. VII, 651 pages. 1979.

Vol. 70: Semantics of Concurrent Computation. Proceedings 1979. Edited by G. Kahn. VI, 368 pages. 1979.

Vol. 71: Automata, Languages and Programming. Proceedings 1979. Edited by H. A. Maurer. IX, 684 pages. 1979.

Vol. 72: Symbolic and Algebraic Computation. Proceedings 1979. Edited by E. W. Ng. XV, 557 pages. 1979.

Vol. 73: Graph-Grammars and Their Application to Computer Science and Biology. Proceedings 1978. Edited by V. Claus, H. Ehrig and G. Rozenberg. VII, 477 pages. 1979.

Vol. 74: Mathematical Foundations of Computer Science. Proceedings 1979. Edited by J. Bečvář. IX, 580 pages. 1979.

Vol. 75: Mathematical Studies of Information Processing. Proceedings 1978. Edited by E. K. Blum, M. Paul and S. Takasu. VIII, 629 pages. 1979.

Vol. 76: Codes for Boundary-Value Problems in Ordinary Differential Equations. Proceedings 1978. Edited by B. Childs et al. VIII, 388 pages. 1979.

Vol. 77: G. V. Bochmann, Architecture of Distributed Computer Systems. VIII, 238 pages. 1979.

Vol. 78: M. Gordon, R. Milner and C. Wadsworth, Edinburgh LCF. VIII, 159 pages. 1979.

Vol. 79: Language Design and Programming Methodology. Proceedings, 1979. Edited by J. Tobias. IX, 255 pages. 1980.

Vol. 80: Pictorial Information Systems. Edited by S. K. Chang and K. S. Fu. IX, 445 pages. 1980.

Vol. 81: Data Base Techniques for Pictorial Applications. Proceedings, 1979. Edited by A. Blaser. XI, 599 pages. 1980.

Vol. 82: J. G. Sanderson, A Relational Theory of Computing. VI, 147 pages. 1980.

Vol. 83: International Symposium Programming. Proceedings, 1980. Edited by B. Robinet. VII, 341 pages. 1980.

Vol. 84: Net Theory and Applications. Proceedings, 1979. Edited by W. Brauer. XIII, 537 Seiten. 1980.

Vol. 85: Automata, Languages and Programming. Proceedings, 1980. Edited by J. de Bakker and J. van Leeuwen. VIII, 671 pages. 1980.

Vol. 86: Abstract Software Specifications. Proceedings, 1979. Edited by D. Bjørner. XIII, 567 pages. 1980

Vol. 87: 5th Conference on Automated Deduction. Proceedings, 1980. Edited by W. Bibel and R. Kowalski. VII, 385 pages. 1980.

Vol. 88: Mathematical Foundations of Computer Science 1980. Proceedings, 1980. Edited by P. Dembiński. VIII, 723 pages. 1980.

Vol. 89: Computer Aided Design - Modelling, Systems Engineering, CAD-Systems. Proceedings, 1980. Edited by J. Encarnacao. XIV, 461 pages. 1980.

Vol. 90: D. M. Sandford, Using Sophisticated Models in Resolution Theorem Proving. XI, 239 pages. 1980

Vol. 91: D. Wood, Grammar and L Forms: An Introduction. IX, 314 pages. 1980.

Vol. 92: R. Milner, A Calculus of Communication Systems. VI, 171 pages. 1980.

Vol. 93: A. Nijholt, Context-Free Grammars: Covers, Normal Forms, and Parsing. VII, 253 pages. 1980.

Vol. 94: Semantics-Directed Compiler Generation. Proceedings, 1980. Edited by N. D. Jones. V, 489 pages. 1980.

Vol. 95: Ch. D. Marlin, Coroutines. XII, 246 pages. 1980.

Vol. 96: J. L. Peterson, Computer Programs for Spelling Correction: VI, 213 pages. 1980.

Vol. 97: S. Osaki and T. Nishio, Reliability Evaluation of Some Fault-Tolerant Computer Architectures. VI, 129 pages. 1980.

Vol. 98: Towards a Formal Description of Ada. Edited by D. Bjørner and O. N. Oest. XIV, 630 pages. 1980.

Vol. 99: I. Guessarian, Algebraic Semantics. XI, 158 pages. 1981.

Vol. 100: Graphtheoretic Concepts in Computer Science. Edited by H. Noltemeier. X, 403 pages. 1981.

Vol. 101: A. Thayse, Boolean Calculus of Differences. VII, 144 pages. 1981.

Vol. 102: J. H. Davenport, On the Integration of Algebraic Functions. 1–197 pages. 1981.

This series reports new developments in computer science research and teaching – quickly, informally and at a high level. The type of material considered for publication includes:

1. Preliminary drafts of original papers and monographs
2. Lectures on a new field or presentations of a new angle in a classical field
3. Seminar work-outs
4. Reports of meetings, provided they are
    a) of exceptional interest and
    b) devoted to a single topic.

Texts which are out of print but still in demand may also be considered if they fall within these categories.

The timeliness of a manuscript is more important than its form, which may be unfinished or tentative. Thus, in some instances, proofs may be merely outlined and results presented which have been or will later be published elsewhere. If possible, a subject index should be included. Publication of Lecture Notes is intended as a service to the international computer science community, in that a commercial publisher, Springer-Verlag, can offer a wide distribution of documents which would otherwise have a restricted readership. Once published and copyrighted, they can be documented in the scientific literature.

Manuscripts

Manuscripts should be no less than 100 and preferably no more than 500 pages in length.
They are reproduced by a photographic process and therefore must be typed with extreme care. Symbols not on the typewriter should be inserted by hand in indelible black ink. Corrections to the typescript should be made by pasting in the new text or painting out errors with white correction fluid. Authors receive 75 free copies and are free to use the material in other publications. The typescript is reduced slightly in size during reproduction; best results will not be obtained unless the text on any one page is kept within the overall limit of 18 x 26.5 cm (7 x 10½ inches). On request, the publisher will supply special paper with the typing area outlined.
Manuscripts should be sent to Prof. G. Goos, Institut für Informatik, Universität Karlsruhe, Zirkel 2, 7500 Karlsruhe/Germany, Prof. J. Hartmanis, Cornell University, Dept. of Computer-Science, Ithaca, NY/USA 14850, or directly to Springer-Verlag Heidelberg.

Springer-Verlag, Heidelberger Platz 3, D-1000 Berlin 33
Springer-Verlag, Neuenheimer Landstraße 28–30, D-6900 Heidelberg 1
Springer-Verlag, 175 Fifth Avenue, New York, NY 10010/USA

ISBN 3-540-10290-6
ISBN 3-540-10290-6